Lecture Notes in Economics and Mathematical Systems 592

Lecture Notes in Economics
and Mathematical Systems

503

Founding Editors:

M. Beckmann
H. P. Künzi

Managing Editors:

Prof. Dr. G. Fandel
Fachbereich Wirtschaftswissenschaften
Fernuniversität Hagen
Feithstr. 140/AVZ II, 58084 Hagen, Germany

Prof. Dr. W. Trockel
Institut für Mathematische Wirtschaftsforschung (IMW)
Universität Bielefeld
Universitätsstr. 25, 33615 Bielefeld, Germany

Editorial Board:

A. Basile, A. Drexl, H. Dawid, K. Inderfurth, W. Kürsten, U. Schittko

Abraham C.-L. Chian

Complex Systems Approach to Economic Dynamics

With 40 Figures

 Springer

Professor Abraham C.-L. Chian
National Institute for Space Research (INPE)
P.O. Box 515
São José dos Campos-SP 12227-010
Brazil
E-mail: achian@dge.inpe.br

Library of Congress Control Number: 2007923717

ISSN 0075-8442

ISBN 978-3-540-39752-6 Springer Berlin Heidelberg New York

Springer is a part of Springer Science+Business Media

springer.com

© Springer-Verlag Berlin Heidelberg 2007

Production: SPS, India
Cover-design: WMX Design GmbH, Heidelberg

SPIN 11856757   88/3100YL - 5 4 3 2 1 0   Printed on acid-free paper

I wish to dedicate this monograph to my parents
Mr. Jong Hong Chian and Mrs. Pi-Sia Wong Chian

Preface

Characterization of the complex dynamics of economic cycles, by identifying regular and irregular patterns and regime switching between different dynamic phases in the economic time series, is the key to improve economic forecasting. Statistical analysis of stock markets and foreign exchange markets have demonstrated the intermittent nature of nonlinear economic time series, which exhibits non-Gaussian behavior in the probability distribution function of price changes and power-law dependence on frequency in the spectral density. Nonlinear deterministic models of economic dynamics are capable of simulating intermittent time series arising from a transition from order to chaos, or from weak chaos to strong chaos, which can explain the origin and nature of intermittency observed in economic systems.

This monograph studies complex economic dynamics based on a forced van der Pol oscillator model of business cycles. The technique of numerical modeling is applied to characterize the fundamental properties of complex economic systems which present multiscale and multistability behaviors, as well as coexistence of order and chaos. In particular, we focus on the dynamics and structure of unstable periodic orbits and chaotic saddles within a periodic window of the bifurcation diagram, at the onset of a saddle-node bifurcation and at the onset of an attractor merging crisis, as well as in the chaotic regions associated with type-I intermittency and crisis-induced intermittency, in nonlinear economic cycles. Inside a periodic window, chaotic saddles are responsible for the transient motion preceding convergence to a periodic attractor or a chaotic attractor. The links between chaotic saddles, crisis and intermittency in complex economic dynamics are discussed. We show that a chaotic attractor is composed of chaotic saddles and unstable periodic orbits located in the gap regions of chaotic saddles. Both type-I intermittency and crisis-induced intermittency are the results of the occurrence of explosion following the onset of a local or a global bifurcation, respectively, whereby the gap regions of chaotic saddles are filled by coupling unstable periodic orbits.

Nonlinear modeling of economic chaotic saddle, crisis and intermittency can improve our understanding of the dynamics of economic intermittency observed in business cycles and financial markets. In view of the universal mathematical nature of chaotic systems, the results obtained from our simple prototype model of economic dynamics can in fact be applied to more complex economic scenarios, including nonlinear spatiotemporal economic systems. Characterization of the complex dynamics of economic systems provides an efficient guide for pattern recognition and forecasting the turning points of business and financial cycles, as well as for optimization of management strategy and decision technology.

I wish to thank Dr. Colin Rogers, Dr. Erico Rempel, Dr. Felix Borotto, Mr. Rodrigo Miranda, and Mr. Wanderson Santana for their collaboration, assistance and friendship. I wish to thank Dr. Tönus Puu and Dr. Steve Keen for their constructive and critical comments. I wish to thank my wife, Kwai Lin, and my daughters, Clarice, Elisa and Janice for their love and prayers.

April 2007   *Abraham C.-L. Chian*
School of Economics,
University of Adelaide, Australia
& National Institute for Space Research, Brazil

Contents

1

Introduction

Economic systems exhibit ubiquitous complex dynamics evidenced by large-amplitude and aperiodic fluctuations in economic variables such as foreign exchange rates, gross domestic product, interest rates, production, stock market prices and unemployment (Hommes 2004). Traditionally, economists have studied economic dynamics using the Newtonian approach by treating the economic fluctuations as linear perturbations near the equilibrium (Scarth 1996, Gandolfo 1997, Shone 2002). The linear approach is valid only for small-amplitude fluctuations and cannot describe the complex characteristics of large-amplitude and aperiodic economic fluctuations. Large-amplitude fluctuations in economic and financial systems are indications that these systems are driven far away from the equilibrium whereby the nonlinearity dominates the system behavior; aperiodic economic and financial fluctuations are manifestations of chaos intrinsic in a complex system. Hence, a non-Newtonian approach based on nonlinear dynamics is required to understand the nature of complex economic dynamics.

In recent years, there is a growing interest in applying nonlinear dynamics to economic modeling. For example, Chiarella (1988) introduced a general nonlinear supply function into the traditional cobweb model under adaptive expectations, and showed that in its locally unstable region it contains a regime of period-doubling followed by a chaotic regime. Puu (1991) studied the nonlinear dynamics of two competing firms in a market in terms of Cournot's duopoly theory; by assuming iso-elastic demand and constant unit production costs this model shows persistent periodic and chaotic motions. Keen (1995) introduced a real financial sector and two stylized facts into Goodwin's growth cycle model; the resulting nonlinear system is able to model the complex behavior of Minsky's financial instability hypothesis, with the transition from stability to instability and possible breakdown determined by the level of economic inequality, interest rate and debt. Scarth (1996) derived a nonlinear standard aggregate demand and supply model of a closed economy consisting of IS, LM, and Phillips curve relationships, described by the logistic function which admits chaotic cycles for a range of control parameters; this model indicates that the standard practice of linear approximations in macroeconomics is a definite limitation. Brock and Hommes (1997) applied the concept of adaptively rational equilibrium to a cobweb type demand-supply model where agents can choose between rational and naive expectations, which shows that in an unstable market with positive information costs for rational expectations, a high intensity of choice to switch predictors leads to highly irregular equilibrium prices converging to complex dynamics such as a strange attractor. Rosser (2001) showed that in an integrated global ecologic-economic system a variety of chaotic and catastrophic patterns appear in the models of global warming dynamics and fishery dynamics, which complicate global policy making efforts. Hughston and Rafailidis (2005) applied a chaotic approach to develop dynamical models for interest rates and foreign exchange; they used the Wiener chaos expansion technique to formulate a systematic analysis of the structure and classification of these financial models. Many more examples of nonlinear economical modeling can be found in the books on complex economic dynamics (Puu 1989, Chiarella 1990, Zhang 1990, Brock, Hsieh and LeBaron 1991, Rosser 1991, Benhabib 1992, Medio 1992, Lorenz 1993, Day 1994, 2000, Thomas, Reitz and Samanidou 2005).

One of the main signatures of complex systems is intermittency, which is characterized by abrupt changes of the system activity with alternating periods of quiescent low-level fluctuations and bursting high-level fluctuations. Temporal intermittency and spatiotemporal intermittent turbulence are pervasive in nature and society, e.g., the flow of cars in heavy traffic in the cities, the floods and droughts of rivers such as the Nile, the fluid turbulence in atmosphere and ocean, and the sunspot cycles (Vassilicos 1995). Intermittency exhibits multiscale behavior (power-law dependence on frequency/wavenumber) and non-Gaussian statistics (heavy-tail probability distribution function of fluctuations), involving information transfer between different scales. There is evidence that intermittency is also a fundamental feature of complex economic and financial systems. For example, Müller et al. (1990) presented a statistical analysis of four foreign exchange spot rates against the U.S. dollar; they found that the mean absolute changes of logarithmic prices follow a scaling law against the interval on which they are measured and there is a net flow of information from long to short timescales, which implies that the behavior of long-term traders (who watch the markets only from time to time) influences the behavior of short-term traders (who watch the markets continuously). Mantegna and Stanley (1995) showed that the scaling of the probability distributions of the Standard & Poor 500 index can be described by a non-Gaussian process with dynamics that, for the central part of the distribution, corresponds to that predicted for a Lévy stable process. Ghashghaie et al. (1996) reported an analogy between the information cascade in foreign exchange market and the energy cascade in hydrodynamic turbulence, and concluded that the intermittent behavior of turbulent flows, with typical occurrence of laminar periods which are interrupted by turbulent bursts, corresponds to clusters of high and low volatility in the foreign exchange markets, which gives rise to relatively high values of the probability densities of price changes both in the center and the tails. Krawiecki, Holyst and Helbing (2002) considered a model of financial markets consisting of many interacting agents, and obtained time series of price returns showing chaotic bursts resulting from the emergence of attractor bubbling or on-off intermittency, resembling the empirical financial time series with volatility clustering; the probability distributions of returns exhibit power-law tails. Mattedi et al. (2004) studied the financial risk of the aerospace sector and developed a new index for this sector based on the New York exchange and the Over the Counter markets. They showed that the statistical characteristics of this index are more volatile but less intermittent than other traditional market indicators such as the Dow-Jones industrial index and the Standard & Poor 500 index. This suggests the existence of long memory correlations having an impact on the volatility clustering patterns of this index.

Chaotic systems are known to describe various types of intermittency, which occur whenever the behavior of a system seems to switch back and forth between two (or more) qualitatively different behaviors even though all the control parameters are kept constant and no noise is present (Hilborn 1994). The intermittent route to chaos was first discovered by Manneville and Pomeau (1979). They identified three types of intermittency whereby the system seems to switch between periodic/quasiperiodic behavior and chaotic behavior due to a transition from order to chaos via a local bifurcation such as saddle-node (tangent) bifurcation, Hopf bifurcation, or period-doubling bifurcation. Another chaotic scenario that leads to intermittency occurs when the system undergoes a global bifurcation known as crisis (Grebogi, Ott and York 1983), whereby a chaotic attractor in the state space suddenly changes in size (interior crisis), disappears (boundary crisis), or two or more chaotic attractors merge to form a large chaotic attractor (attractor merging crisis). In crisis-induced intermittency, the systems switch between weakly chaotic and strongly chaotic behaviors (Grebogi, Ott and Romeiras 1987). There are many examples of experimental observations of chaos-driven intermittency. For example, Hayashi, Ishizuka and Hirakawa (1983) observed a transition from order to chaos via type-I Pomeau-Manneville intermittency in the onchidium pacemaker neuron. Ditto et al. (1989) observed crisis-induced intermittency in a magnetoelastic ribbon experiment.

Stable and unstable periodic orbits are the basic elements of complex dynamical systems, and are the key to explain the origin and nature of chaos-driven intermittency. A complex system consists of order and chaos; order is governed by stable periodic orbits, whereas chaos is governed by unstable periodic orbits. In particular, unstable periodic orbits are the skeleton of chaotic attractors and chaotic saddles (Auerbach et al. 1987, Cvitanovic 1988, Hilborn 1994). Chaotic saddles are non-attracting chaotic sets which are responsible for chaotic transients (Grebogi, Ott and Yorke 1983, Kantz and Grassberger 1985), and are the backbones of chaotic attractors (Szabó and Tél, 1994a, 1994b). In addition, chaotic saddles are responsible for intermittency in the chaotic regions outside a periodic window (Szabó et al. 2000), e.g., beyond a saddle-node bifurcation (type-I intermittency), and beyond an interior crisis (crisis-induced intermittency). There is experimental evidence of unstable periodic orbits, chaotic transients and chaotic saddles. For example, Schief et al. (1994) detected the presence of unstable fixed-point behavior in a spontaneously bursting neuronal network in vitro and demonstrated that chaos in brain dynamics can be controlled and anticontrolled by changing the stability properties of the unstable fixed point. Jánosi, Flepp and Tél (1994) reconstructed the chaotic transient behavior of a laser based on a long time series in a laboratory experiment; they showed that the motion on the chaotic transient is more unstable than on the coexisting chaotic attractor. Faisst and

Eckhardt (2003) identified a family of unstable traveling waves, originating from saddle-node bifurcations, in a numerical experiment for flow through a pipe; these unstable structures provide a skeleton for the formation of a chaotic saddle that can explain the intermittent transition to turbulence and the sensitive dependence on initial conditions in this flow.

Chaotic transients and chaotic saddles are fundamental to the understanding of complex economic dynamics. Lorenz (1993) observed chaotic transient motion in a Kaldorian model of business cycles. Lorenz and Nusse (2002) demonstrated the potential relevance of chaotic saddles in the Goodwin's nonlinear accelerator model of business cycles. Apart from the works by Lorenz (1993) and Lorenz and Nusse (2002), most economic literature and books on complex economic dynamics (Puu 1989, Chiarella 1990, Zhang 1990, Benhabib 1992, Brock, Hsieh and LeBaron 1991, Rosser 1991, Medio 1992, Day 1994, 2000, Thomas, Reitz and Samanidou 2005) have only dealt with chaotic attractors, paying no attention to chaotic transients and chaotic saddles.

In Chapter 2, a forced van der Pol oscillator model of economic cycles is formulated as the prototype model to describe the complex economic dynamics. The fundamental properties of nonlinear dynamics of economic cycles are studied, including discussions on order and chaos, Poincaré map, bifurcation diagram and periodic window, multistablilty and basins of attraction, unstable periodic orbit and chaotic attractor.

In Chapter 3, based on numerical simulations of the forced oscillator model of nonlinear economic cyles, it is shown that after an economic system undergoes a dynamical transition from an ordered to a chaotic state, the system maintains its memory before the transition and the economic variables switch alternatively between periods of quiescent and bursting fluctuations. This type-I economic intermittency arises from a local bifurcation known as the saddle-node bifurcation. An economic path evolves from a periodic to an aperiodic pattern when the exogenous forcing amplitude passes a critical value whereby the system loses its stability due to a saddle-node bifurcation. The power spectrum of the type-I intermittent time series is broadband and displays power-law behavior at high frequencies, similar to the real data of foreign exchange and stock markets. The characteristic intermittency time, measuring the average duration of quiescent periods in the intermittent economic time series, is a function of the exogenous forcing amplitude. The scaling law of the characteristic intermittency time is useful for forecasting the turning points of nonlinear economic cycles.

In Chapter 4, a new type of crisis-induced intermittency in nonlinear economic cycles is discussed. It is shown that after an economic system undergoes a global bifurcation known as attractor merging crisis, the system has the ability to keep the memory of its weakly chaotic state before crisis. As the result, the economic variables switch alternatively between periods of weakly and strongly chaotic fluctuations. Similar to the type-I economic intermittency, the power spectrum of the time series of the crisis-induced economic intermittency is broadband and presents power-law behavior at high frequencies, typical of volatile financial markets. As the system moves away from the crisis point, it becomes more chaotic, consequently the discrete spikes of the power spectrum become less evident due to increasing multiscale information transfer in the complex economic systems. The exponent of the scaling law of the characteristic intermittency time of the crisis-induced economic intermittency is much larger than that of the type-I economic intermittency.

In Chapter 5, an attractor merging crisis in chaotic economic cycles is characterized. It is shown that the van der Pol model of economic cycles is invariant under the flip operation. Symmetry is a common property of complex systems that exhibit attractor merging crisis. The analysis is performed in a complex region within a periodic window of the bifurcation diagram determined from the numerical solutions of a forced oscillator, where a saddle-node bifurcation marks the beginning of the periodic window. As the exogenous forcing amplitude increases after the saddle-node bifurcation, two coexisting periodic attractors of period-1 undergo a cascade of period-doubling bifurcations leading to two weakly chaotic attractors. An attractor merging crisis occurs when two coexisting weakly chaotic attractors merge to form a single strongly chaotic attractor, which marks the end of the periodic window. The onset of attractor merging crisis is due to the head-on collision of the pair of coexisting weakly chaotic attractors, respectively, with a pair of mediating unstable periodic orbits of period-3 and their associated stable manifolds. In addition, it is demonstrated that the two coexisting weakly chaotic attractors also collide with the boundary of the basins of attraction that separates the two weakly chaotic attractors.

The aim of Chapter 6 is to perform an in-depth study of unstable periodic orbits and chaotic saddles in complex economic dynamics. In particular, the roles of unstable periodic orbits and chaotic saddles in crisis and intermittency in complex economic systems are investigated. The technique of numerical modeling is applied to characterize the dynamics and structure of unstable periodic orbits and chaotic saddles within a periodic window of the bifurcation diagram, at the onset of a saddle-node bifurcation and of an attractor merging crisis, as well as in type-I intermittency and crisis-induced intermittency, of a forced oscillator model of economic cycles. The links between chaotic saddles, crisis and intermittency in complex economic dynamics are analyzed.

The conclusion is given in Chapter 7.

2

Nonlinear Dynamics of Economic Cycles

Complex dynamics of economic systems can be studied by applying the concepts and techniques of nonlinear dynamics and chaos. Some models of business cycles, such as Kaldor's nonlinear investment-savings functions and Goodwin's nonlinear accelerator-multiplier, can be reduced to the van der Pol equation which describes relaxation oscillations. By introducing an exogenous driver, the forced van der Pol equation can be adopted as a prototype model for complex economic dynamics. Numerical solutions of this model can elucidate the fundamental properties of complex economic systems which exhibit a wealth of nonlinear behaviors such as multistability as well as coexistence of order and chaos. Unstable periodic orbits are the skeleton of chaotic attractors in complex economic systems.

2.1 Empirical Evidence of Nonlinearity and Chaos in Economic Data

Recently, there is a growing interest in nonlinear dynamics and chaos in economics. Actual economic time series are rarely characterized by regular (periodic, sinusoidal) dynamics typical of linear systems. Instead, various types of irregular (aperiodic, non-sinusoidal) forms of large-amplitude fluctuations in economic time series are often observed, which cannot be adequately explained by linear analysis. The significant fluctuations indicated by many economic variables relative to their mean values suggest that most economic systems are far away from the equilibrium, i.e., inherently nonlinear.

Chaotic motions can arise in nonlinear economic systems if the time series is aperiodic and displays sensitive dependence on initial conditions (Puu 1989, Lorenz 1993). Empirical evidence of complex behaviors of nonlinear deterministic systems can be obtained by calculating statistical quantities such as Lyapunov exponents, entropies, fractal dimensions, and correlation dimensions. These quantitative measures of chaos are defined for infinitely large data sets. In practice, large amount of data points are often unavailable in macroeconomic time series. In contrast to the laboratory experiments where a large amount of data points can easily be obtained, most economic time series consists of monthly, quarterly, or annual data, with the exception of some financial data with daily or weekly time series. This imposes severe limitation on the accuracy of nonlinear analysis of economic data. In view of this limitation, additional tests are desirable.

Brock (1986) performed a test for chaos in detrended quarterly US real GNP data from 1947 to 1985 by calculating the correlation dimension and the largest Lyapunov exponent and applying an additional residual test, and concluded that chaos should be excluded in the GNP data. Barnett and Chen (1988) examined several monetary aggregates and found positive values for the largest Lyapunov exponents in some of their data, which provides evidence of chaos. Frank, Gencay and Stengos (1988) applied the shuffle test proposed by Scheinkman and LeBaron (1989) to German, Italian, and U. K. GNP data, and ruled out the presence of chaos in their GNP data, but found evidence of nonlinearity. Sayers (1989) calculated the correlation dimension and the Lyapunov exponents and applied the additional residual diagnostics to U. S. business cycles, including GNP, pig-iron production, and unemployment rates. Although he did not find the presence of chaos, he obtained evidence of nonlinear structures. Further literature survey on empirical evidence of nonlinearity and chaos in economical data will be given in the remaining chapters of this monograph.

2.2 Modeling Nonlinearity and Chaos in Economic Dynamics

Nonlinear dynamics models are useful to explain irregular, large-amplitude, fluctuations that appear in complex economic systems (Hommes 2004). The complex behaviors of nonlinear economic systems restrict the use of purely analytical methods to investigate nonlinear economic models. In general, numerical simulations provide the most efficient way to derive information from nonlinear economic models. In contrast to nonlinear analysis of economic data which are restricted by the small sample size as well as noise, numerical modeling of economic systems can provide large sample size required to characterize chaotic behaviors, and determine the dynamical behaviors of economic systems in the absence and in the presence of noise. Economic models can be formulated by either discrete-time or continuous-time approaches (Puu 1989; Lorenz 1993).

From the outset (Samuelson 1939; Hicks 1950), business cycle models have most frequently been formulated in discrete time, as difference equations or iterated maps such as the logistic map (Scarth 1996). The main reason for taking the discrete-time approach is the relative facility to handle these models, without the need of heaving computation. For example, Stutzer (1980) characterized the qualitative dynamics of a discrete-time version of a nonlinear macroeconomic model, which shows complex periodic and random aperiodic orbit structures. Nusse and Hommes (1990) considered a discrete modified Samuelson model of nonlinear multiplier-accelerator and showed that period-doubling bifurcation and period-halving bifurcation leading to chaos can occur; the chaos disappears when the accelerator is increased. Day and Pavlov (2002) developed a variation of Goodwin's graphical model to explain the rudiments of Keynesian real/monetary cycle theory, which possesses nonlinear dynamical properties of irregular, asymmetric fluctuations. Xu et al. (2002) studied the Kaldor's business cycle model in two-dimensional discrete form and introduced an approach to detect cyclical patterns (unstable periodic orbits) embedded in chaotic economic data and made use of the detected patterns to estimate the trends of periodic-like motions in a chaotic evolution of economic systems.

A large number of nonlinear business cycle models are formulated in continuous time based on either ordinary or partial differential equations. New econometric techniques emerged recently permit direct empirical testing of continuous-time economic models. In this monograph, the continuous-time approach will be adopted. Goodwin (1951) was one of the first Keynesian economists to introduce nonlinear continuous-time dynamical model with locally unstable steady states and stable limit cycles to account for the persistence of business long wave which explains the Kondratieff economic cycle in terms of subsequent expansions and contractions of the capital goods sector of an industrialized economy, as it adjusts to the required production capacity.

Lorenz (1987a) studied a continuous-time model of three coupled sectors of Kaldor-type business cycles, and showed that if the sectors are linked by investment demand interdependencies, this coupling can be interpreted as a perturbation of a motion on a three-dimensional torus; chaotic fluctuations appear in this model. Sasakura (1995) investigated political business cycles in two different forced oscillator models of the Duffin-type and van de Pol-type, respectively, by incorporating autonomous investment and Kaldor-type induced investment function; in both cases chaotic fluctuations emerge even when the politically motivated fiscal forcing is weak. Additional literature survey on nonlinear economic models will be discussed in the remaining chapters of this monograph.

2.3 Van der Pol Model of Nonlinear Business Cycles: Kaldor's Nonlinear Investiment-Savings Functions

Inspired by Keyne's income theory and Kalecki's model of investiment (Kalecki 1937), Kaldor (1940) formulated the first nonlinear model of endogenous business cycles by considering the interactions between the investiment $I(Y)$ and the savings $S(Y)$ functions (where Y denotes income) and the existence conditions for self-sustaining limite cycles. By noting that the linear forms of $I(Y)$ and $S(Y)$ fail to produce cyclical motions, Kaldor proposed a S-shaped (sigmoid) nonlinear form for $I(Y)$ and a mirror-imaged S-shaped nonlinear form for $S(Y)$ (Gabisch and Lorenz 1989), which yields oscillatory motion of business cycles. Chang and Smyth (1971) reformulated Kaldor's model of business cycles, given by the following couped dynamical equations:

$$\dot{Y} = \alpha(I(Y,K) - S(Y,K)), \qquad (2.1)$$
$$\dot{K} = I(Y,K), \qquad (2.2)$$

where the dot denotes the time derivative (d/dt), K denotes the capital stock and α is an adjustment coefficient; with the assumptions of $I_K < 0$, $S_K > 0$, and $\partial S/\partial K < 0$.

Differentiating equation (2.1) with respect to time gives

$$\ddot{Y} = \alpha(I_Y\dot{Y} + I_K\dot{K} - S_Y\dot{Y} - S_K\dot{K}). \qquad (2.3)$$

A substitution of equation (2.2) into equation (2.3) yields

$$\ddot{Y} - \alpha(I_Y - S_Y)\dot{Y} - \alpha(I_K - S_K)I(Y,K) = 0. \qquad (2.4)$$

By treating the investment I_Y as usual, but the actual change in capital stock K is determined by savings, $\dot{K} = S$; and assuming $(I_K - S_K)$ is independent of capital stock so the functions are linear in K, equation (2.4) becomes

$$\ddot{Y} - \alpha(I_Y - S_Y)\dot{Y} - \alpha I_K S(Y) = 0, \qquad (2.5)$$

which reduces to the generalized Liénard equation in physical systems

$$\ddot{x} + A(x)\dot{x} + B(x) = 0, \qquad (2.6)$$

which describes the dynamics of a spring mass system with $A(x)\dot{x}$ as a damping factor and $A(x)$ as the spring force.

By postulating symmetric shapes of the investment and savings functions, and a parabolic functional form for the difference $S_Y - I_Y$, namely,

$$A(x) = \alpha(S_Y - I_Y) = \mu(x^2 - 1), \qquad (2.7)$$

and

$$B(x) = x, \qquad (2.8)$$

equation (2.5) can be rewritten as

$$\ddot{x} + \mu(x^2 - 1)\dot{x} + x = 0, \qquad (2.9)$$

which is known as the Van der Pol equation originally derived by Van der Pol and Van der Mark (1928) to describe relaxation oscillations in an electrical circuit model of the heartbeat, and can serve as a prototype continuous-time model of complex economic dynamics. Note that the parameter μ is related to the adjustment coefficient α of the damping term.

2.4 Forced Van der Pol Model of Nonlinear Economic Cycles: Goodwin's Nonlinear Accelerator-Multiplier with Lagged Investiment Outlays

The concept of business cycles was introduced by Samuelson (1939) by combining the accelerator and the multiplier. This model demostrates that two simple forces related to the producers keeping a fixed ratio of capital stock to output (real income) and the cousumers spending of a given fraction of their incomes on consumption can combine to generate business cycles.

Goodwin (1951) formulated a nonlinear model of business cycles which provides an alternative to the restrictive linear accelerator-multiplier models of Samnelson-Hicks. In contrast to Hicks' model (Hicks 1950) of nonlinear business cycles which assumed that the unconstrained linear accelerator-multiplier model takes on special parameter values that the system will explode, the Goodwin's model does not depend on specific parameter values. By introducing lagged investment outlays in the nonlinear accelerator-multiplier model, Goodwin (1951) derived the following driven oscillator equation

$$\epsilon\theta\ddot{y} + (\epsilon + (1 - \alpha)\theta)\dot{y} - \phi(\dot{y}) + (1 - \alpha)y = I(t), \qquad (2.10)$$

where y denotes income, α is the marginal rate of consumption, ϵ is a constant denoting a lag in the dynamic multiplier process, θ is the lag between the decision to invest and the corresponding outlays, $\phi(\dot{y})$ is the investment induced by the change in income, and $I(t)$ is an exogenous force denoting the amount of autonomous investiment outlays at t.

Lorentz (1987b) and Lorenz and Nusse (2002) considered the following generalization of equation (2.10) to consider chaotic motion in Goodwin's nonlinear accelerator-multiplier

$$\ddot{x} + A(x)\dot{x} + B(x) = I(t), \qquad (2.11)$$

where $A(x)$ is an even function with $A(0) < 0$, and $B(x)$ is an odd function with $B(0) = 0$. By assuming the investment outlays is periodic and continuous function of time

$$I(t) = a\sin(\omega t), \qquad (2.12)$$

where a is the amplitude of exogenous force and ω the frequency of exogenous force, and

$$A(x) = \mu(x^2 - 1), B(x) = x, \qquad (2.13)$$

we obtain a forced van der Pol model of nonlinear economic cycles

$$\ddot{x} + \mu(x^2 - 1)\dot{x} + x = a\sin(\omega t). \qquad (2.14)$$

In addition to Kaldor's nonlinear investment-savings functions and Goodwin's nonlinear accelerator-multiplier, the forced Van der Pol model of relaxation oscillations, given by equation (2.14), have many other relevant economical applications (Gabisch nd Lorenz 1989; Puu

1989; Goodwin 1990; Medio 1992; Lorenz 1993; Gandolfo 1997; Shone 2002; Chian 2001; Chian et al. 2005, 2006; Chian, Rempel and Rogers 2006, 2007). The modern economy consists of a great variety of separate sectors and activities closely coupled to each other. For example, Puu (1989) showed that the forced Van der Pol equation similar to equation (2.14) can model the nonlinear dynamics of a small economy driven exogenously by the the world market, which can produce very rich dynamical solutions including a chaotic atractor. Puu's model of international trade provides an illustration of the interdependence of an individual national economy and the world economy. Cyclic flutuations are common characteristics of economic systems. A variety of economic cyclic modes have been identified, including the 3-7 year business cycle, the 15-25 year construction or Kuznets cycle, and the 40-60 Kondratieff or economic long wave. Nonlinear interaction between different economic modes can occur, e.g., a short-period business cycle can act as an exogenous force on a long-period business cycle. In addition, geophyical cycles such as seasonal cycles, El Niño cycles, and solar cycles may act as an exogenous force on the fluctutations of the agriculture, tourism, and fuel sectors.

In this monograph, we will investigate the numerical solutions of the forced van der Pol model of business cycles, equation (2.14), which can be rewritten as three coupled first-order differential equations

$$\dot{x}_1 = x_2, \tag{2.15}$$

$$\dot{x}_2 = -\mu(x_1^2 - 1)x_2 - x_1 + a\sin(2\pi x_3), \tag{2.16}$$

$$\dot{x}_3 = \frac{\omega}{2\pi}. \tag{2.17}$$

In the absence of the exogenous forcing ($a = 0$), the origin ($x_1 = 0$, $x_2 = 0$) is the only equilibrium solution of equations (2.15)-(2.17), which is an unstable fixed point (repeller); all other trajectories of the system approach a single attracting periodic orbit (limit cycle) that encircles the origin, which describes periodic relaxation oscillations consisted of a period of slow buildup followed by a sudden discharge (Alligood, Sauer and Yorke 1996). In the presence of an exogenous forcing, equation (2.14) admits a rich variety of periodic and aperiodic oscillations as the control parameters μ, a and ω are varied. Parlitz and Lauterborn (1987) gave examples of the bifurcation diagrams of equation (2.14) by varying the driver frequency and the driver amplitude, which show mode-locking and period-doubling cascade. They pointed out that the system symmetry of the van der Pol oscillator leads to the coexistence of asymmetric attractors, and introduced a generalized winding number to compute devil's staircases and winding-number diagrams of period-doubling cascades. For large driving amplitudes, they found that many periodic, quasiperiodic and chaotic attractors coexist. A systematic analysis of equation (2.14) was carried out by Mettin, Parlitz and Lauterborn (1993) by studying its dynamical behaviors over a large range of control parameters in the three-dimensional (μ, a, ω) phase diagrams, paying special attention to the pattern of the bifurcation curves in the transitional region between low and large dampings. Xu and Jiang (1996) performed a global bifurcation analysis of equation (2.14) by investigating the phase diagrams in the two-dimensional (μ, a) plane with a fixed ω, for medium damping. They studied the evolution of the global structures in simple and complex transitional zones, and the number of coexisting attractors in overlaps of mode-locking subzones.

In this chapter, we use the numerical solutions of equation (2.14) to study the fundamental properties of nonlinear dynamics of economic cycles.

2.5 Order and Chaos

One fundamental characteristic of a complex dynamical system is the possibility of order and chaos, which can exist either separately or simultaneously. In an ordered dynamical system, for arbitrary initial conditions, after going through a transient period the system approaches a periodic behavior with a predictable periodicity. In contrast, a chaotic dynamical system exhibits behavior that depends sensitively on the initial conditions, thereby rendering long-term prediction impossible (Strogatz 1994). Figure 2.1(a) shows a periodic time series of the numerical solutions of equation (2.14) for the control parameters: $\mu = 1$, $\omega = 0.45$, $a = 0.983139$. Figure 2.1(b) shows two chaotic time series of the numerical solutions of equation (2.14) for the same control parameters: $\mu = 1$, $\omega = 0.45$ and $a = 0.9877$, but with two slightly different set of initial conditions. The initial conditions of the solid curve are $x = 0.2108$, $\dot{x} = 0.0187$; whereas, the initial conditions of the dashed curve are $x = 0.2100$, $\dot{x} = 0.0187$. We see from figure 2.1(b) that initially the two time series are the same, however, as time increases, the behavior of the two chaotic time series becomes very different.

The attractor is the set of points in the state space to which the trajectories approach as time goes to infinity. Since a complex system

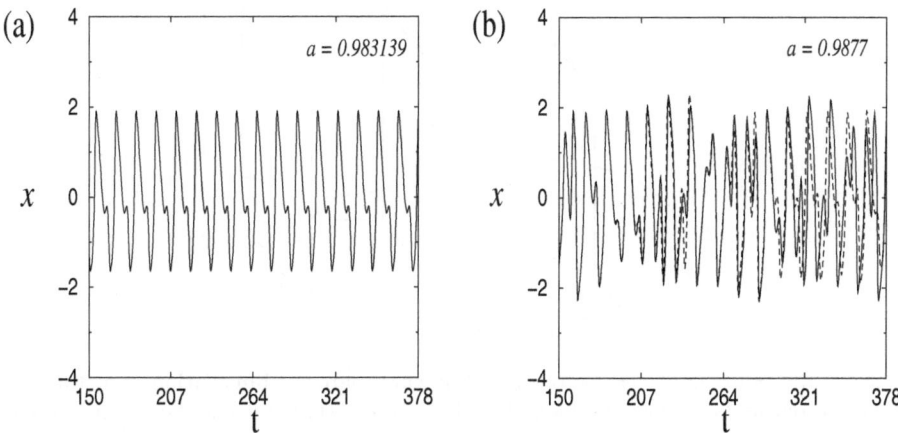

Fig. 2.1. Periodic and chaotic time series. (a) A periodic time series $x(t)$ for $a = 0.983139$, (b) two chaotic time series for $a = 0.9877$ with slightly different initial conditions: $x = 0.2108$ and $\dot{x} = 0.0187$ for the solid line, $x = 0.2100$ and $\dot{x} = 0.0187$ for the dashed line.

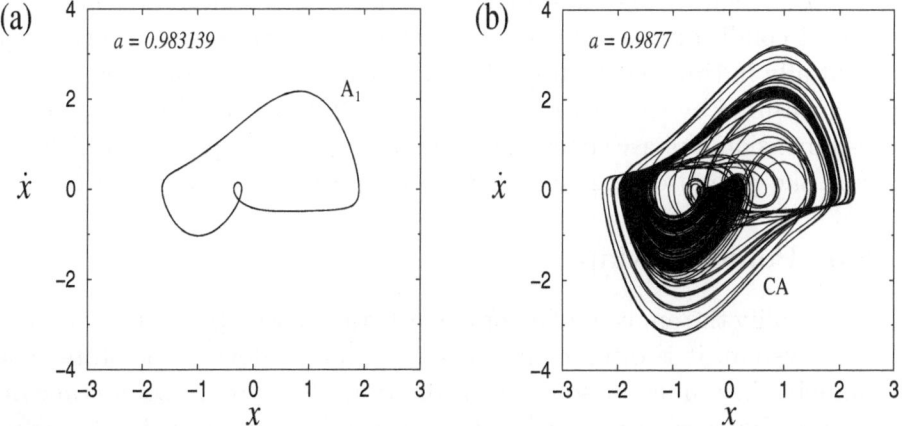

Fig. 2.2. Periodic attractor and chaotic attractor. (a) A periodic attractor (A_1) of period-1 in the state space $(x, \dot{x})$ for $a = 0.983139$, (b) a chaotic attractor (CA) in the state space $(x, \dot{x})$ for $a = 0.9877$.

consists of both order and chaos, it contains both periodic attractors and chaotic attractors. When the attractor is an isolated closed trajectory, it is called a periodic attractor (or limit cycle); when an attractor is a fractal set of points, it is called a strange attractor (or chaotic attractor) (Ott 1993). Figures 2.2(a) gives an example of a periodic attractor

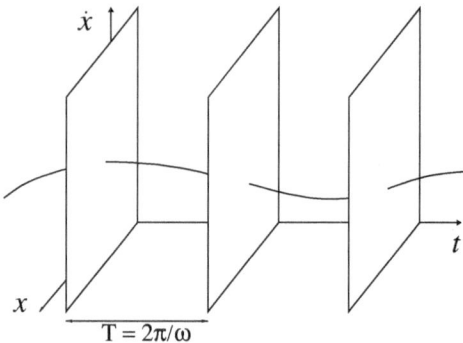

Fig. 2.3. State-space trajectory and Poincaré map. An illustration of a state-space trajectory and the Poincaré map, T is the driver period and ω is the driver frequency.

(A_1) for $a = 0.983139$, corresponding to the periodic time series in figure 2.1(a). Figure 2.2(b) gives an example of a chaotic attractor for $a = 0.9877$ (CA), corresponding to the chaotic time series in figure 2.1(b). The trajectories of arbitrary initial conditions on a chaotic attractor will display aperiodic behavior and sensitive dependence on initial conditions, which implies that nearby orbits will diverge exponentially in time (see figure 2.1(b)). The average rate of divergence can be measured by the Lyapunov exponents (Ott 1993). For a system with n-dimensional phase space, there are n Lyapunov exponents which measure the rate of divergence/convergence in n orthogonal directions.

2.6 Poincaré Map

To simplify the analysis of a nonlinear trajectory (orbit or flow) of a complex system, it is often convenient to reduce a flow in the state space, namely, the numerical solution of equation (2.14), to a discrete time map by the Poincaré surface of section method (Ott 1993). In this monograph, we define the Poincaré surface of section (Poincaré map) by

$$P : x(t) \rightarrow x(t + T), \tag{2.18}$$

where $T = 2\pi/\omega$ is the driver period. Figure 2.3 is an illustration of a state-space trajectory and the Poincaré map.

2.7 Bifurcation Diagram and Periodic Window

In addition to sensitive dependence on the initial conditions, a dynamical system is very sensitive to small variations in the control parameters

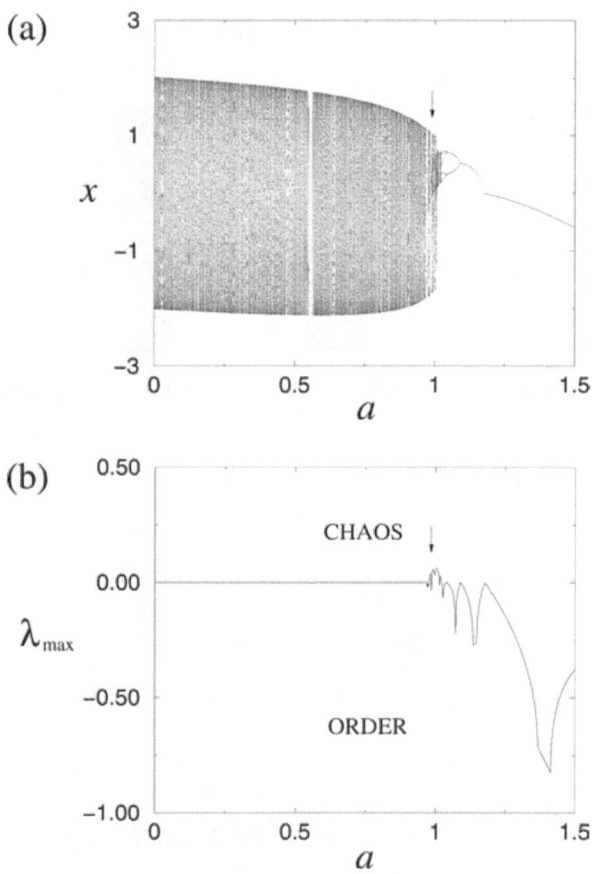

Fig. 2.4. Bifurcation diagram and maximum Lyapunov exponent: global view. (a) Bifurcation diagram, x as a function of a, (b) the maximum Lyapunov exponent λ_{max} as a function of a, positive λ_{max} indicates chaos and negative λ_{max} indicates order; with fixed $\mu = 1$ and $\omega = 0.45$.

(either endogenous or exogenous). As a control parameter varies, the stability of a dynamical system changes due to a local or a global bifurcation. The bifurcation diagram provides a general view of the system dynamics by plotting a system variable as a function of a control parameter (Alligood, Sauer and Yorke 1996). Figure 2.4(a) shows a global view of the bifurcation diagram of the nonlinear model of economic cycles described by equation (2.14), where we have kept two control parameter μ and ω fixed, and only vary the forcing amplitude a. For a given control parameter a, the bifurcation diagram in figure 2.4(a) plots

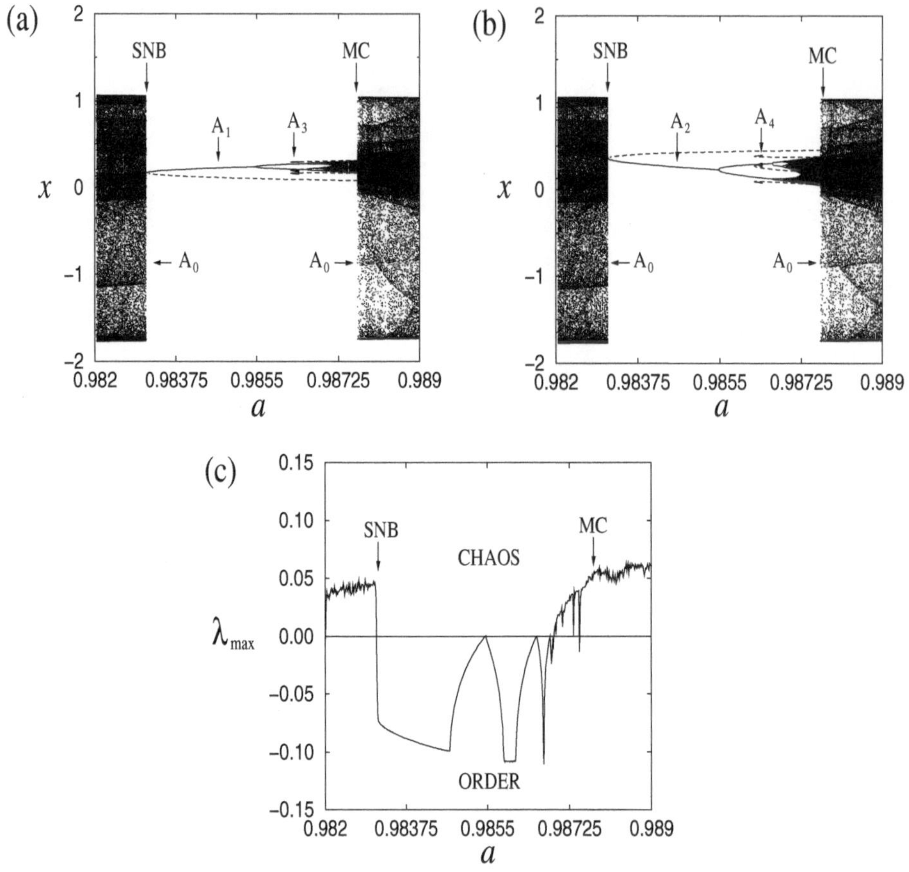

Fig. 2.5. Bifurcation diagram and maximum Lyapunov exponent: periodic window. (a) Bifurcation diagram, x as a function of a, for attractors A_0, A_1 and A_3; (b) bifurcation diagram for attractors A_0, A_2 and A_4; (c) the maximum Lyapunov exponent λ_{max} as a function of a for attractors A_0, A_1 and A_2. SNB denotes saddle-node bifurcation, MC denotes attractor merging crisis.

the asymptotic values of the Poincaré points of the system variable x, where the transient has been omitted.

The phase space of equations (2.15)-(2.17) has three dimensions, therefore the system has three Lyapunov exponents, one of which is always zero (in the direction tangent to the flow). For the remaining two exponents, for a stable periodic orbit the maximum Lyapunov exponent is less than zero, for a quasiperiodic orbit the maximum Lyapunov exponent is zero, whereas for a chaotic orbit the maximum Lyapunov exponent is greater than zero. Figure 2.4(b) shows the maximum Lyapunov exponent as a function of a, for the bifurcation diagram given by figure 2.4(a), calculated by the Wolf algorithm (Wolf et al. 1985). Figure 2.4 shows that the system is quasiperiodic to the left of $a \sim 1$, and periodic to the right of $a \sim 1$. However, in the region $a \sim 1$, the system can be chaotic.

An enlargement of a small region of the bifurcation diagram indicated by the arrow in figure 2.4(a) is given in figures 2.5(a) and 2.5(b), which display a periodic window. Complex dynamics is found within this periodic window, where five attractors are identified. A saddle-node bifurcation (SNB) at $a = a_{SNB} = 0.98312$ marks the beginning of this periodic window (in terms of attractors A_1 and A_2). An attractor merging crisis (MC) at $a = a_{MC} = 0.98765$ marks the end of this periodic window. To the left of a_{SNB} and to the right of a_{MC}, we have a chaotic attractor A_0. Two attractors A_1 and A_2 co-exist between a_{SNB} and a_{MC}, throughout this periodic window. Two more attractors A_3 and A_4 coexist for a small interval of a, between $a = 0.9862400$ and $a = 0.9864085$. Due to the symmetry of equation (2.1) the attractors A_1 and A_2 have the same dynamical behaviors, namely, for a given control parameter a, the maximum Lyapunov exponents of A_1 and A_2 are the same. The same is true for attractors A_3 and A_4. Figure 2.5(c) shows the maximum Lyapunov exponent for either attractor A_1 or attractor A_2, which indicates that there are many small periodic windows within a chaotic region, and there are many chaotic regions within a periodic window. The rich dynamics found in this periodic window demonstrates the basic features of multistability and coexistence of order and chaos in complex economic systems. In this monograph, we focus on the periodic window given by figure 2.5 to investigate the complex dynamical behaviors of economic systems.

2.8 Multistability and Basins of Attraction

Evidently, multistability is a fundamental feature of a complex system, as seen in the periodic window of the bifurcation diagram in figures 2.5(a) and 2.5(b). The basin of attraction for a given attractor is the set of initial conditions each of which gives rise to a trajectory that converges asymptotically to the attractor (Hilborn 1994). Note that the chaotic attractor A_0 persists to the right of $a_{SNB} = 0.983120$ and is only destroyed by a boundary crisis at $a = 0.983139$. In terms of attractor A_0, the periodic window actually starts at $a = 0.983140$. Hence, three attractors A_0, A_1 and A_2 coexist between $a = 0.983120$ and $a = 0.983139$, as exemplified by the basins of attraction in figure 2.6(a) for $a = 0.983139$. For the initial conditions starting from the light gray

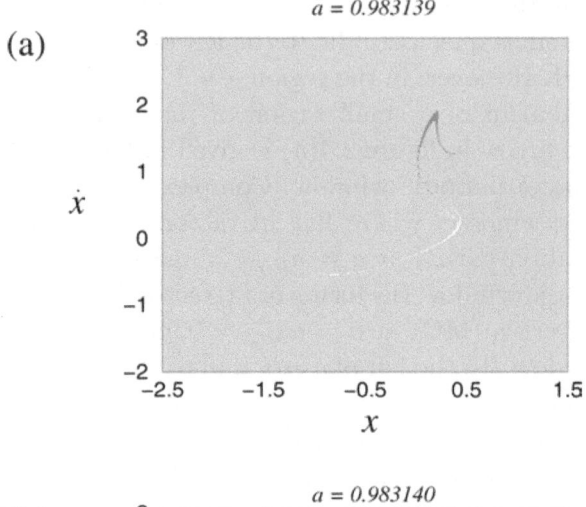

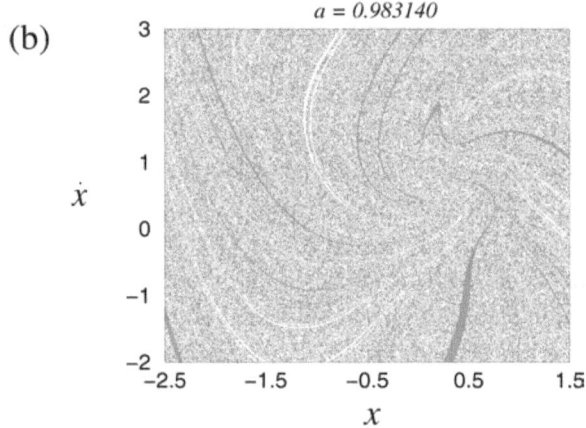

Fig. 2.6. Basins of attraction: multistability. Basins of attraction for: (a) $a = 983139$ with coexistence of three attractors A_0, A_1 and A_2; (b) $a = 0.983140$ with coexistence of two attractors A_1 and A_2. Attractor A_0 (light gray), attractor A_1 (dark gray), attractor A_2 (white).

region, the trajectory converges to the chaotic attractor A_0; whereas, for initial conditions starting in the dark gray (white) region, the trajectory converges to the periodic attractor A_1 (A_2), respectively. Between $a = 0.983140$ and $a = 0.9862399$, and between $a = 0.9864086$ and $a = a_{MC} = 0.98765$, two attractors A_1 and A_2 coexist, as exemplified by the basins of attraction in figure 2.6(b) for $a = 0.983140$, where the light gray (white) region denotes the basin of attraction for attractor A_1 (A_2). Note the dramatic change in the topology of the basins of attraction in figures 2.6(a) and 2.6(b), where the control parameter varies slightly from $a = 0.983139$ to $a = 0.983140$. This dramatic change is due to the destruction of the chaotic attractor A_0 and its basin of attraction by a boundary crisis. Four attractors A_1, A_2, A_3 and A_4 coexist between $a = 0.9862400$ and $a = 0.9864085$.

2.9 Unstable Periodic Orbit and Chaotic Attractor

Unstable periodic orbits are the skeleton of a chaotic attractor because chaotic trajectories are closures of the set of unstable periodic orbits (Auerbach et al. 1987, Cvitanovic 1988). In contrast to a periodic attractor whereby all trajectories initiated from any point in the state space are attracted to a stable periodic orbit (e.g., figure 2.2(a)), in a chaotic attractor all periodic orbits are unstable since almost all trajectories (with the exception of trajectories strictly along its stable manifold) in the neighborhood of an unstable periodic orbit are repelled by it (e.g., figure 2.2(b)). Hence, a chaotic trajectory is chaotic because it must weave in and around all of these unstable periodic orbits yet remain in a bounded region of state space (Hilborn 1994). Unstable periodic orbits can be numerically found by the Newton algorithm (Curry 1979). Four examples of the state-space trajectory (solid line) and Poincaré points (cross) of unstable periodic orbits are given in figure 2.7. The saddle-node bifurcation at $a = a_{SNB} = 0.98312$ generates a pair of stable and unstable periodic orbits of period-1 associated with attractors A_1 and A_2, respectively, as shown in figures 2.7(a) and 2.7(b). Note that the stable and unstable periodic orbits are identical at the onset of a saddle-node bifurcation. These two period-1 unstable periodic orbits, represented by the dashed lines to the right of $a = 0.98312$ in figures 2.5(a) and 2.5(b), are responsible for mediating the onset of a boundary crisis at $a = 0.983139$ which destroys the chaotic attractor A_0. Figures 2.7(c) and 2.7(d) show the unstable (stable) periodic orbits of period-3 associated with attractors A_3 and A_4, respectively, generated by another saddle-node bifurcation at $a = 0.9862400$. These two period-3 unstable periodic orbits, represented by the dashed lines to the right of $a = 0.9862400$ in figure 2.5(a), are responsible for mediating the onset of another boundary crisis that destroys attractors A_3 and A_4 at $a = 0.9864085$, and are also responsible for mediating the onset of an attractor merging crisis (MC) at $a = 0.98765$, which marks the end of the periodic window in figure 2.5. The unstable periodic orbits are robust. For example, most unstable periodic orbits that appear within the periodic window continue to exist in the chaotic region to

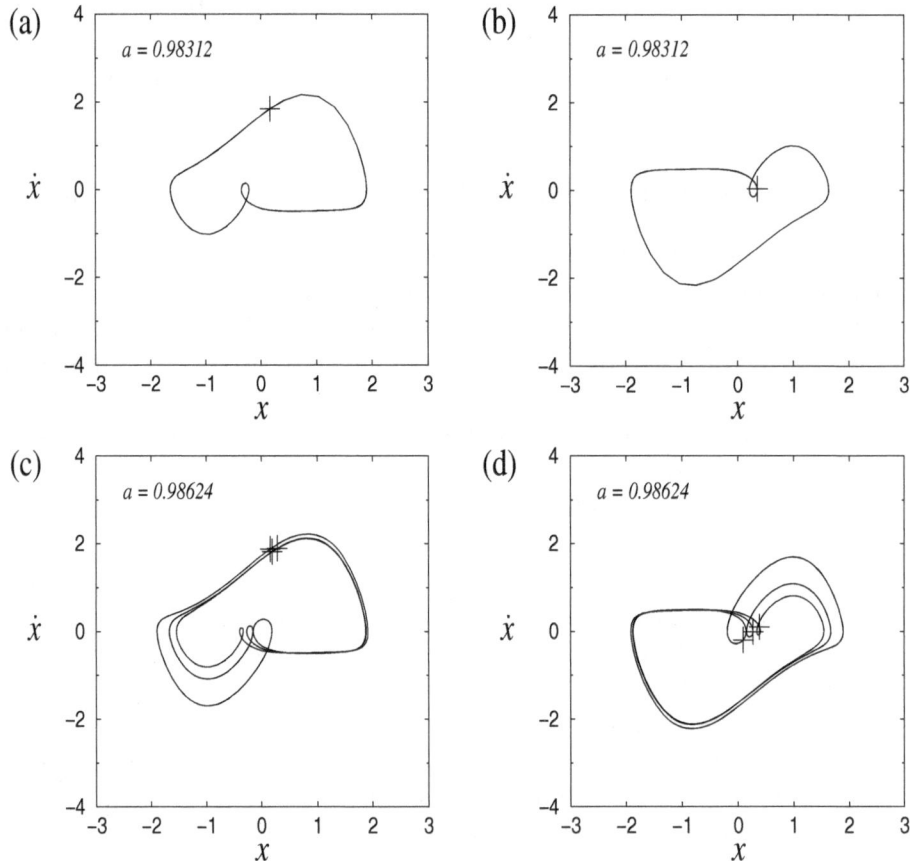

Fig. 2.7. Unstable periodic orbits: skeleton of chaotic attractor. Examples of unstable periodic orbits (solid line) in the state space and the corresponding Poincaré point (cross) of: (a) and (b) period-1 for $a = 0.98312$, (c) and (d) period-3 for $a = 0.98624$.

the right of MC in figure 2.5(a) and form part of the skeleton of the chaotic attractor A_0 beyond the attractor merging crisis.

An unstable periodic orbit with period-N turns into N-saddle points in the Poincaré surface of section, as seen in Figure 2.7. Figure 2.8(a) illustrates a saddle point (p), which is the intersection of in-set (stable manifold SM) and out-set (unstable manifold UM), in a two-dimensional Poincaré surface of section. The dashed lines represent the stable (v^s) and unstable (v^u) eigenvectors of the linearized Poincaré map at p. At the saddle fixed point p, the stable manifold SM is tangent to the stable eigenvector v^s and the unstable manifold UM is tangent to the unstable

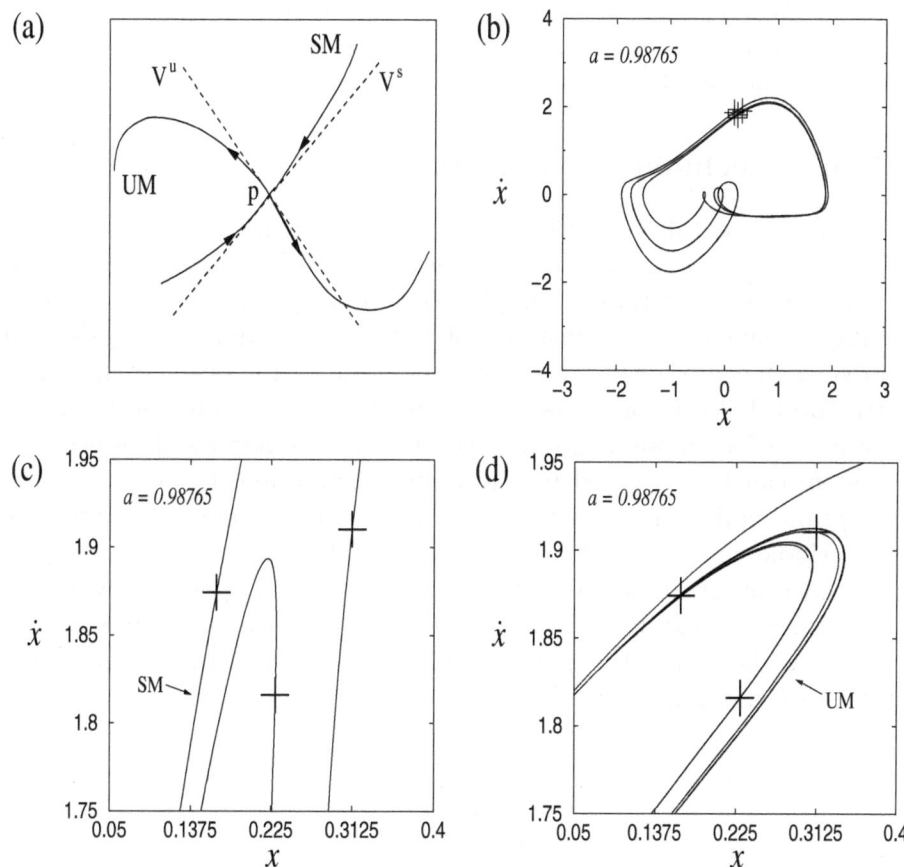

Fig. 2.8. Unstable periodic orbit: stable and unstable manifolds. (a) An illustration of a saddle point (p) with its associated stable manifold (SM) and unstable manifold (UM), the dashed lines represent the stable (v^s) and unstable (v^u) eigenvectors; (b) the state-space trajectory (solid line) and Poincaré points (cross) of a period-3 unstable periodic orbit for $a = a_{MC} = 0.98765$; (c) the stable manifold SM (line) of the period-3 saddle point (cross); (d) the unstable manifold UM (line) of the period-3 saddle point (cross).

eigenvector v^u. Trajectories on the in-set converge to the saddle point as time goes on, while trajectories on the out-set diverge from the saddle point as time goes on (Hilborn 1994). Figure 2.8(b) is an example of the trajectory of an unstable periodic orbit of period-3 in the state space for $a = a_{MC} = 0.98765$. The closed curve in figure 2.8(b) turns into a saddle consisting of 3 fixed points (crosses) in the Poincaré surface of section also shown in figure 2.8(b). Figures 2.8(c) and 2.8(d) are enlargements of the rectangular region indicated in figure 2.8(b), where we also plotted the numerically computed stable manifold (SM) and unstable manifold (UM) of the saddle, respectively.

2.10 Concluding Comments

The fundamental properties of nonlinear economic dynamics discussed in this chapter form the basis for the analysis of complex economic systems. We showed that a complex economic system exhibits multistability behavior with coexistence of attractors, including the possibility of coexistence of order and chaos (periodic attractors and chaotic attractors). In addition, we showed that unstable periodic orbits are the skeleton of a chaotic attractor. The complex dynamics of an economic system can be displayed by the Poincaré map and by the bifurcation diagram, which often contains many periodic windows. We identified a periodic window within which complex dynamics is found, with the presence of five attractors; the beginning of this periodic window is marked by a saddle-node bifurcation (in terms of attractors A_1 and A_2) and the end of this periodic window is marked by the onset of an attractor merging crisis.

3

Type-I Intermittency in Nonlinear Economic Cycles

In this chapter, the intermittent behavior of economic dynamics is studied by a nonlinear model of business cycles. Numerical simulations show that after an economic system evolves from order to chaos, the system keeps its memory before the transition and its time series alternates episodically between periods of low-level apparently periodic (quiescent) and high-level turbulent (bursting) activities. This model of economic intermittency exhibits power-law spectrum similar to the nonlinear time series observed in financial markets.

3.1 Introduction

Characterization of the complex dynamics of economic cycles, by organizing economic regularities and identifying regime switching between "good" and "bad" phases in the time series, is the key to accurate economic forecasting (Diebold and Rudebusch 1999). In a classical book, Burns and Mitchell (1946) defined business cycles as "a type of fluctuation found in the aggregate economic activity of nations that organize their work mainly in business enterprises: a cycle consists of expansions, occurring at about the same time in many economic activities, followed by similarly general recessions, contractions, and revivals which merge into the expansion phase of the next cycles". Thus, two fundamental attributes of business cycles are: comovement (i.e., synchronization) among various economic variables or sectors, and division of business cycles into alternating (i.e., intermittent) phases of low-level and high-level economic activities.

Synchronization and intermittency are ubiquitous phenomena that govern the nonlinear dynamics of complex systems. Fireflies provide a good example of synchronization in nature whereby thousands of fireflies can self-organize themselves to flash on and off in synchrony. Periodic (ordered) solutions appear when coupled oscillators are phase-locked due to phase synchronization; moreover, phase synchronization can occur in coupled chaotic oscillators (Strogatz 1994). Selover et al. (2004) proposed that national business cycles result from nonlinear phase-locking between different industries or sectors. Intermittency is pervasive in our world, as exemplified by traffic flow in big cities, fluid turbulence in atmospheres and oceans, and long-term variabilities of sunspot cycles (Vassilicos 1995; Ossendrijver and Covas 2003). Financial markets also exhibit intermittent behavior wherein periods of trading frenzy are followed by periods of quiescence; on closer examination the periods of high volatility are themselves consisted of other sub-periods of relative quiet and other sub-periods of relative bursty activities, which is a manifestation of self-similar and scale-invariant properties of nonlinear systems.

Recent statistical analysis of high-frequency data of stock markets and foreign exchange markets have demonstrated the intermittent nature of nonlinear economic time series, which present non-Gaussian behavior in the probability distribution function of price changes and power-law behavior in the spectral density (Mantegna and Stanley 1995, 1996; Ghashghaie et al. 1996). The fat-tail seen in the non-Gaussian probability distribution function is due to excess of large-amplitude fluctuations (relative to Gaussian distribution) of economic variables. The power-law frequency dependence of the spectral density is an indication of turbulent process involving an information cascade from large to small time scales in financial markets.

There is an increasing interest in applying chaos concept to study nonlinear economic dynamics. Sengupta and Sfeir (1997) performed empirical tests of volatility for monthly data of exchange rates from February 1988 to August 1995, and concluded that chaotic instability cannot be ruled out in general. Fernandez-Rodriguez et al. (1997) applied a multivariate local predictor, inspired by chaos theory, to nine EMS currencies using daily data from January 1973 to December 1994, which outperformed the random walk directional forecasting. Muckley (2004) presented evidence of strange attractor, a long-term memory effect and aperiodic motion in a time series analysis of daily financial data of two equity and two commodity indices.

Intermittency is readily found in nonlinear models of economic dynamics (Mosekilde et al. 1992; Haxholdt et al. 1995; Bischi et al. 1998). In this chapter, we study an example of economic type-I intermittency based on a model of nonlinear business cycles (Chian et al. 2006). We will show by numerical simulations that after a transition from order to chaos due to a saddle-node bifurcation, the time series of business cycles becomes intermittent involving episodic regime switching between quiescent and bursting phases. The power spectrum of the simulated intermittent time series has power-law dependence on frequency, similar to the observed data of intermittent financial markets. The characteristic intermittency time will be calculated and its application for economic forecasting will be discussed.

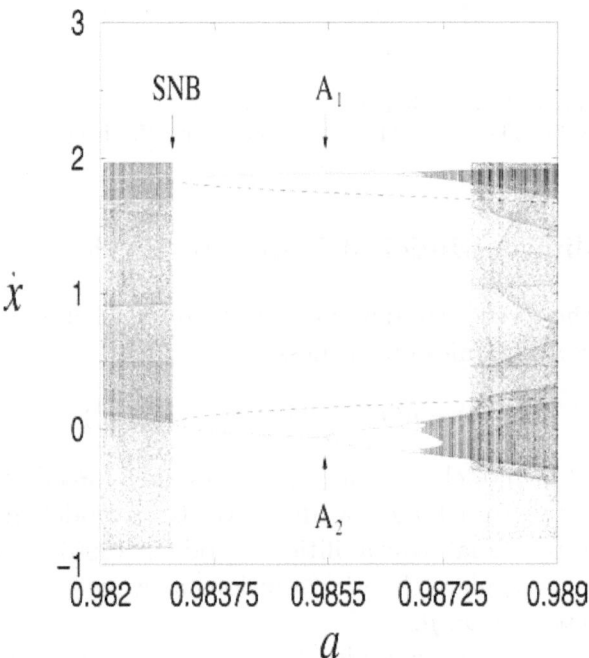

Fig. 3.1. A p-1 periodic window of computed bifurcation diagram, $\dot{x}$ as a function of the driver amplitude a, for attractors A_1 and A_2. SNB denotes saddle-node bifurcation; dashed lines denote p-1 unstable periodic orbits; $\mu = 1$ and $\omega = 0.45$.

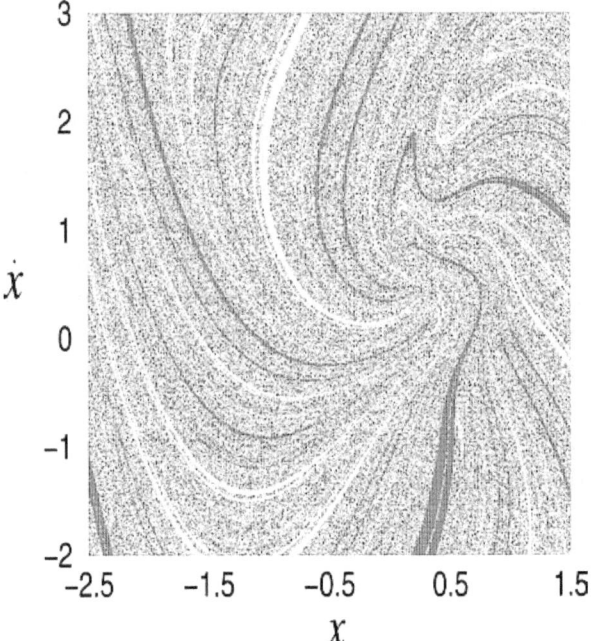

Fig. 3.2. Basins of attraction for two co-existing attractors A_1 and A_2 at $a = a_{SNB} = 0.98314$. The gray (white) regions denote the basins of attraction of A_1 (A_2).

3.2   Nonlinear Model of Economic Cycles

We adopt the forced van der Pol (VDP) differential equation to model the nonlinear dynamics of business cycles

$$\ddot{x} + \mu(x^2 - 1)\dot{x} + x = a\sin(\omega t). \tag{3.1}$$

Equation (3.1) models a small open economy forced externally by a world economy (Puu 1989), or alternatively, it models market fluctuations driven by climate variabilities (Goodwin 1990). It admits regular (periodic) or irregular (chaotic) solutions as we vary any of three control parameters: a, ω, μ.

Equation 3.1 is an example of two coupled oscillators: an endogenous nonlinear oscillator with its natural frequency, and an exogenous periodic oscillator with a driver frequency ω. In a nonlinear system, the natural frequency of oscillations changes with the variation of the control parameters. Hence, in this economic model the dynamical behavior of nonlinear business cycles depends on the competition between

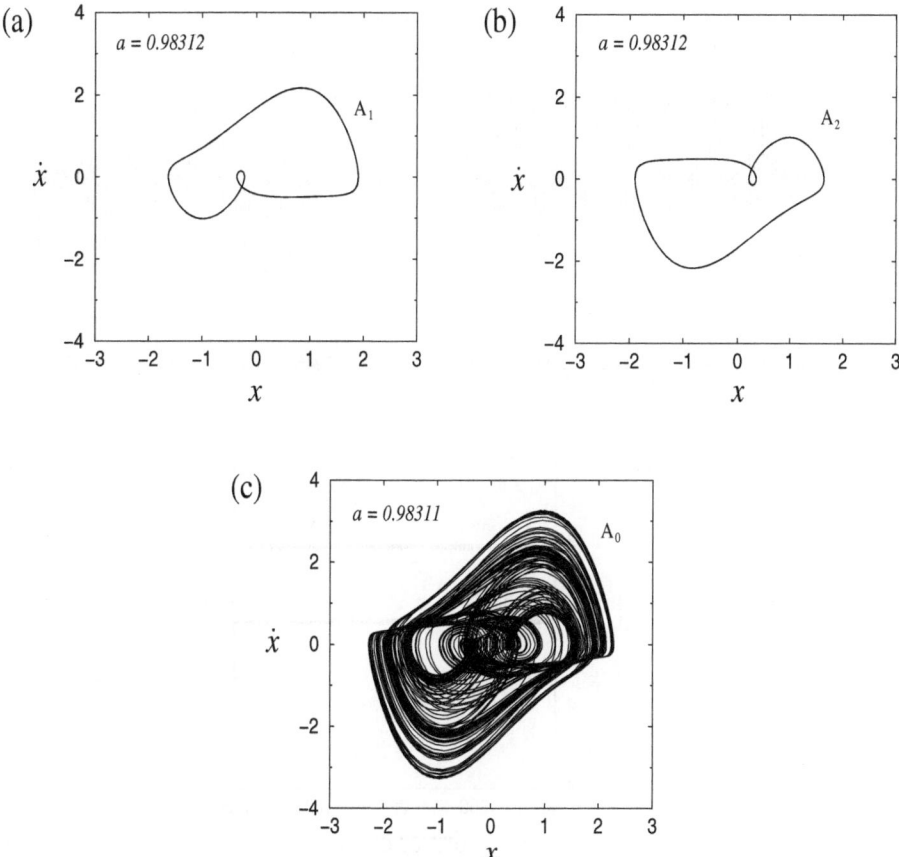

Fig. 3.3. Phase-space trajectories of: (a) period-1 attractor (A$_1$) for $a = 0.98312$, (b) period-1 attractor (A$_2$) for $a = 0.98312$, (c) chaotic attractor (A$_0$) for $a = 0.98311$

these two frequencies as the control parameters are varied. The system is phase-locked (synchronized) if the ratio of these two frequencies is a rational number; its associated solution is then periodic. After the phase-locked solution is destroyed in a saddle-node bifurcation, the solution becomes chaotic. Type-I intermittency results from the transition from order to chaos via a saddle-node bifurcation (Strogatz 1994).

3.3 Economic Type-I Intermittency

A periodic window of the bifurcation diagram determined from the numerical solutions of equation (3.1) is shown in figure 3.1, where we plot

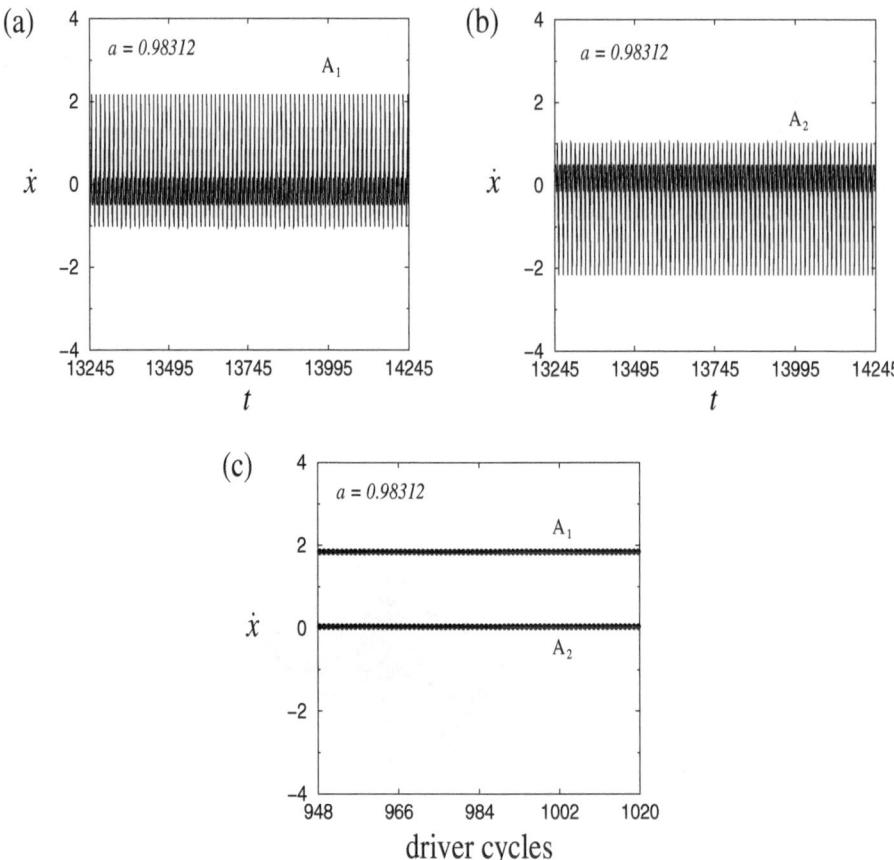

Fig. 3.4. Periodic time series for $a = a_{SNB} = 0.98312$: (a) $\dot{x}(t)$ for attractor A_1, (b) $\dot{x}(t)$ attractor A_2, (c) $\dot{x}$ as a function of driver cycles for A_1 and A_2

$\dot{x}$ as a function of the amplitude a of the exogenous forcing while keeping other control parameters fixed ($\mu = 1$ and $\omega = 0.45$) (Chian et al. 2006). Within the periodic window, two (or more) coexisting attractors A_1 and A_2 are found. At the saddle-node bifurcation $a = a_{SNB} = 0.98312$, a pair of period-1 (p-1) stable (solid line) and unstable (dashed line) periodic orbits for each attractor is generated, which evolve into two small chaotic attractors via a cascade of period-doubling bifurcations. To the left of a_{SNB} in the bifurcation diagram, the initial conditions converge to a chaotic attractor A_0. The aim of this chapter is to study type-I intermittency associated with the transition of periodic attractors A_1/A_2 to the chaotic attractor A_0 for $a < a_{SNB}$.

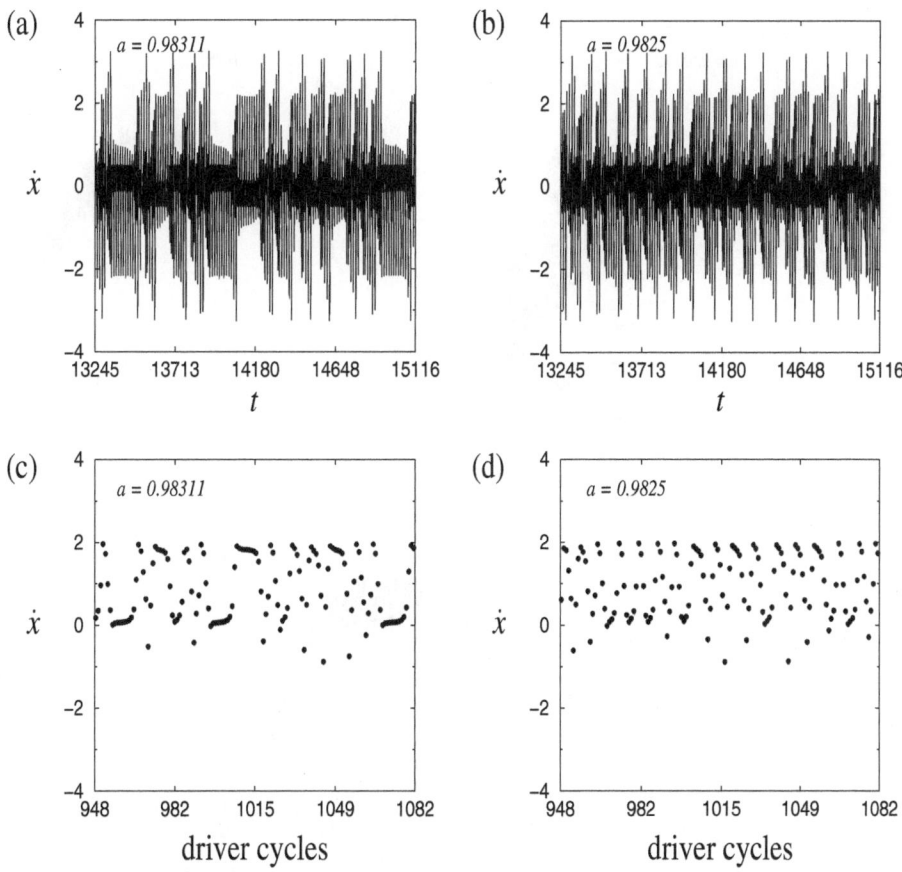

Fig. 3.5. Intermittent time series for $a = 0.98311$ and $a = 0.9825$. (a) and (b): $\dot{x}(t)$, (c) and (d): $\dot{x}$ as a function of driver cycles.

Due to the symmetry of equation (3.1), which is invariant under the flip operation $x \to -x$ when $a = 0$, the solutions admit coexistence of attractors. Figure 3.2 shows the basins of attraction for attractors A_1 and A_2 at $a = 0.98314$, within the periodic window. The set of initial conditions in the gray region of the phase space $(x, \dot{x})$ will be attracted to A_1, whereas the set of initial conditions in the white region will be attracted to A_2. Note that for values of a between 0.983120 and 0.983139 the three attractors A_1, A_2 and A_0 coexist. The chaotic attractor A_0 is destroyed by a boundary crisis (BC) at $a_{BC} = 0.983139$, to the right of a_{SNB}.

At $a = a_{SNB}$, the attractors A_1 and A_2 are periodic with period-1. The trajectories of A_1 and A_2 in the phase space $(x, \dot{x})$ at $a = a_{SNB}$ are

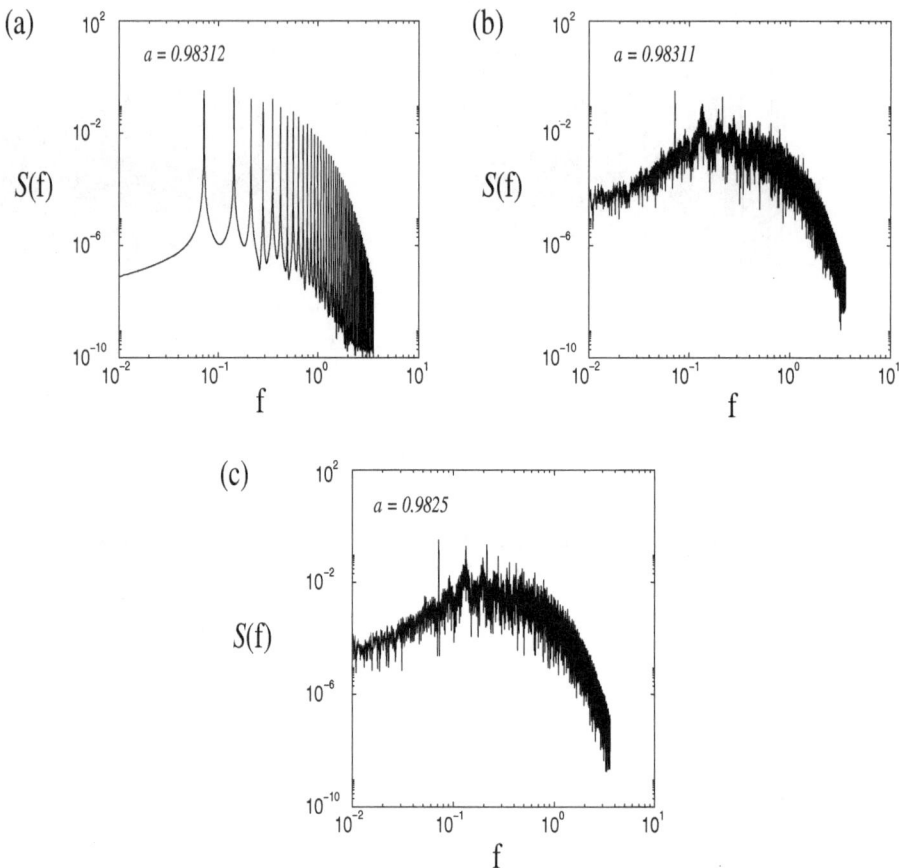

Fig. 3.6. Power spectrum $S(f)$ as a function of frequency f for: (a) $a = 0.98312$, (b) $a = 0.98311$, (c) $a = 0.9825$

shown in figures 3.3(a) and 3.3(b), respectively. Examples of periodic time series, $\dot{x}(t)$, for attractors A_1 and A_2 are shown in figures 3.4(a) and 3.4(b), respectively; the same time series plotted as a function of driver cycles ($t = 2\pi n/\omega$, $n = 1, 2, 3, ...$) are given in figure 3.4(c).

For $a < a_{SNB}$, the solutions are chaotic. The phase-space trajectory of the chaotic attractor A_0 prior to the saddle-node bifurcation is shown in figure 3.3(c). Two examples of chaotic time series for different values of a, to the left of $a = a_{SNB}$, are shown in figures 3.5(a) and 3.5(b), respectively; the same time series plotted as a function of driver cycles are given in figures 3.5(c) and 3.5(d), respectively. Type-I intermittency is readily recognized in Fig. 3.5, exhibiting episodic regime switching

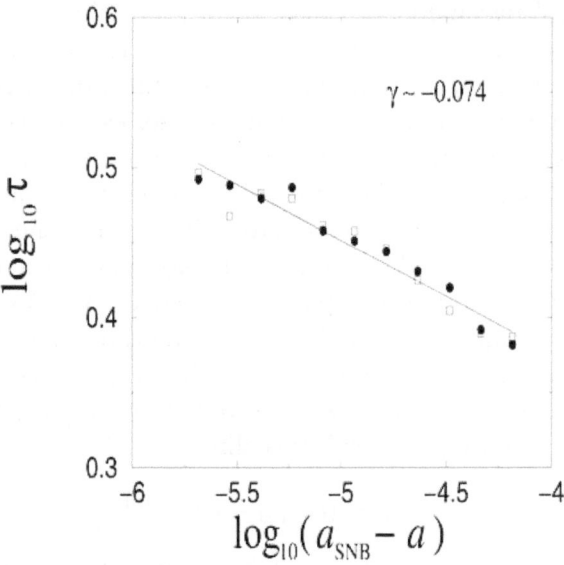

Fig. 3.7. Characteristic intermittency time τ as a function of the departure from a_{SNB}, $\log_{10}\tau$ as a function of $\log_{10}(a_{SNB} - a)$

between periods of laminar (quiescent) phases and periods of bursting (turbulent) phases. By comparing figures 3.4 and 3.5, we identify the laminar phases ($\dot{x} \sim 2$ and $\dot{x} \sim 0$ in the driver cycle plots) as due to the memory effect of the post saddle-node bifurcation p-1 unstable periodic orbits of A_1 and A_2, respectively. As the system moves farther away from the transition point $a = a_{SNB}$, the average duration of laminar phases decreases due to weakening memory, as shown by the intermittent time series in figure 3.5. This implies that after the transition from order to chaos, the regime switching of intermittent business cycles becomes more frequent as the system moves farther away from the transition point.

The power spectra of the periodic and intermittent time series of figures 3.4 and 3.5 are shown in figure 3.6. Figure 3.6(a) shows that when the solution is periodic the spectrum is discrete. Figures 3.6(b) and 3.6(c) show that when the solutions are intermittent the power spectra are broadband and have a power-law behavior at high frequencies, which is a characteristic of chaotic systems such as intermittent financial markets.

The characteristic intermittency time, namely, the average duration of laminar phases in the intermittent time series, depends on the value of the control parameter a. Close to the transition point a_{SNB} the average duration of laminar phases is relatively longer, and decreases as a moves away from a_{SNB}. The characteristic intermittency time (denoted by τ) can be calculated from a long time series, by averaging the time between two consecutive bursting phases. Figure 3.7 is a plot of $\log_{10} \tau$ versus $\log_{10}(a_{SNB}-a)$, where the solid line with a slope $\gamma = -0.074$ is a linear fit of the values of the characteristic intermittency time computed from the time series. The squares (circles) denote the computed average duration of the laminar phases related to A_1 (A_2). Note that the circles and the squares coincide most of the time, due to the symmetry of A_1 and A_2. Figure 3.7 reveals that the characteristic intermittency time τ decreases with the distance from the critical parameter a_{SNB}, obeying the following power-law scaling:

$$\tau \sim (a_{SNB} - a)^{-0.074}. \tag{3.2}$$

This scaling formula can be used to predict the turning points, from contraction to expansion phases, of nonlinear business cycles.

3.4 Concluding Comments

This chapter shows that after an economic system undergoes a dynamical transition from an ordered to chaotic state, intermittency appears whereby the economic activities switch episodically back and forth between periods of quiescent and bursting fluctuations. As an economic system moves farther away from the transition point, the average duration of quiescent periods decreases. In order to understand the nature of economic intermittent behaviors, we performed a study of type-I intermittency in a nonlinear model of business cycles. In this example of intermittency, an economic path evolves from a regular (periodic) to an irregular (chaotic) pattern as the exogenous forcing amplitude a passes a critical value a_{SNB}, where the system loses its stability due to a saddle-node bifurcation. It is worth emphasizing that there is a region with intermittent chaos for attractor A_0 to the right of a_{SNB} in figure 3.1, for values of a between 0.983120 and 0.983139, which will be a subject of further investigation.

The accuracy of business cycle forecasting relies on a precise estimate of the durations of economic expansions and contractions and of the turning points in business cycles (Vilasuso 1996; Schnader and Stekler 1998; Diebold and Rudebusch 1999). Nonlinear modeling of economic systems provides a powerful tool to simulate regime switching between contraction and expansion phases, and to predict the turning points. In particular, the average duration of quiescent phases in business cycles can be determined from the characteristic intermittency time of the simulated time series. Hence, the dynamical systems approach is extremely useful to analyze patterns in the fluctuations of complex economic systems and valuable for sound policy making.

Some interesting connections can be made between our results and other papers discussed in the present work. For example, Vilasuso (1996) employed nonparametric turning-point tests to investigate the duration of economic expansions and contractions in the United States, which indicated evidence of a turning point to longer expansions in 1929. Our work adopted a nonlinear model of business cycles to simulate the duration of expansions and contractions of an open economy driven by a global market, which can be used to predict the turning point to a long period of economic expansions of a nation, such as detected by Vilasuso (1996). Moreover, type-I intermittency studied in this chapter demonstrates the ability of a chaotic enonomic system to retain the memory of its system dynamics in the ordered regime. When the system is close to its transition point from order to chaos, it keeps this memory for a long duration in the form of quiescent phases in economic fluctuations. This result is in agreement with the nonlinear time-series analysis of financial data performed recently by Muckley (2004), which obtained evidence of a long-term memory effect in a strange attractor.

4

Crisis-Induced Intermittency in Nonlinear Economic Cycles

In this chapter, a new type of economic intermittency is found in nonlinear business cycles. Following a merging crisis, a complex economic system has the ability to retain memory of its weakly chaotic dynamics prior to crisis. The resulting time series exhibits episodic regime switching between periods of weakly and strongly chaotic fluctuations of economic variables. The characteristic intermittency time, useful for forecasting the average duration of contractionary phases and the turning point to the expansionary phase of business cycles, is computed from the simulated time series.

4.1 Introduction

Intermittency is a fundamental dynamical feature of complex economic systems. An intermittent economic time series is characterized by recurrence of regime switching between periods of bursts of high-level fluctuations of economic activities and periods of low-level fluctuations. For example, an instability of the financial system leads to speculative booms followed by subsequent financial crises manifested by violent price movements in financial markets; the recurrence of these events results in business cycles with alternating periods of boom and depression (Mullineux 1990). The spectral density of intermittent economic time series indicates power-law behavior typical of mutiscale systems. Statistical analysis of the high-frequency dynamics of stock markets and foreign exchange markets has proven the intermittent nature of these financial systems, which display non-Gaussian form with fat-tail in the probability distribution function of price changes (Mantegna and Stanley 2000).

A good understanding of regime switching and memory of economic time series is essential for pattern recognition and forecasting of business cycles. Kirikos (2000) compared a random walk with Markov switching-regime processes in forecasting foreign exchange rates; the results suggested that the availability of more past information may be useful in forecasting future exchange rates. Kholodilin (2003) introduced structural shifts in the US composite economic indicator via deterministic dummies and evaluated the US monthly macroeconomic series specified by the regime-switching model. Bautista (2003) used regime-switching-ARCH regression on the Philippine stock market data to estimate its conditional variance and relate to episodes of high volatility including the 1997 Asian financial crisis; this study identified a period of high stock return volatility preceding a bust cycle marked by a sequence of low-growth periods. Granger and Ding (1996) defined long memory as a time series having a slowly declining correlogram, which is a property of fractional integrated processes as well as a number of other processes including nonlinear models; the relevance of long memory is illustrated using absolute returns from a daily stock market index. Resende and Teixeira (2002) assessed long-memory patterns in the Brazilian stock market index (Ibovespa) for periods before and after the Real Stabilization Plan, and obtained evidence of short memory for both periods. Gil-Alana (2004) presented evidence of memory in the dynamics of the real exchange rates in Europe using the fractional integration techniques. Muckley (2004) employed rescaled-range analysis, correlation dimension test and BDS test to obtain evidence of long-memory effect and chaos in daily time series of financial data.

Intermittency is ubiquitous in chaotic economic systems. In a nonlinear macroeconomic model (Mosekilde et al. 1992) describing an economic long wave (or Kondratiev cycle) forced by an exogenous short-term construction (or Kuznets) business cycle represented by a sinusoidal fluctuation in the demand for capital to the goods sector, a chaotic transition known as crisis involving a sudden expansion of chaotic attractor and a complex form of chaos arising from intermittency are observed. In a disaggregated economic long wave model describing two coupled industries (Haxholdt et al. 1995), one representing production of plant and long-lived infrastructure and the other representing short-lived equipment and machinery, mode-locking, quasiperiodic behavior, chaos and intermittency are detected. In a model of an economic duopoly game (Bischi et al. 1998), the phenomenon of synchronization of a two-dimensional discrete dynamical system is studied and on-off intermittency due to a transverse instability is detected.

An example of type-I intermittency in nonlinear business cycles was studied recently (Chian et al. 2006). In the economic type-I intermittency, the recurrence of regime switching between bursty and laminar phases indicates that a nonlinear economic system is capable of keeping the memory of its ordered dynamics after the system evolves from order to chaos due to a local saddle-node bifurcation. Most econometric studies of long memory treat economic data as stochastic processes (Granger and Ding 1996; Resende and Teixeira 2002; Gil-Alana 2004), however real economic systems are a mixture of stochastic and deterministic processes. In this chapter, we adopt the deterministic approach to study a new type of economic intermittency induced by an attractor merging crisis due to a global bifurcation (Chian et al. 2005). We will show that following the onset of an attractor merging crisis, the economic system retains its memory of the weakly chaotic dynamics before the crisis; as a result, the time series of business cycles becomes intermittent displaying episodic regime switching between periods of weakly and strongly chaotic fluctuations.

A forced model of nonlinear business cycles is formulated in Section 4.2. Economic crisis-induced intermittency is analyzed in Section 4.3. Concluding comments are given in Section 4.4.

4.2 Nonlinear Model of Economic Cycles

We model the nonlinear dynamics of business cycles driven by the forced van der Pol differential equation

$$\ddot{x} + \mu(x^2 - 1)\dot{x} + x = a\sin(\omega t). \tag{4.1}$$

Equation (4.1) admits periodic (ordered) or aperiodic (chaotic) solutions as we vary any of three control parameters: a, ω, μ. Equation (4.1) (when $a = 0$) is invariant under the flip operation $(x \rightarrow -x)$. This symmetry is a typical property of dynamical systems that exhibit attractor merging crises (Chian et al. 2005, 2006).

4.3 Economic Crisis-Induced Intermittency

The qualitative structure of the trajectory described by equation (4.1) can change (i.e., bifurcate) as the control parameters are varied. For

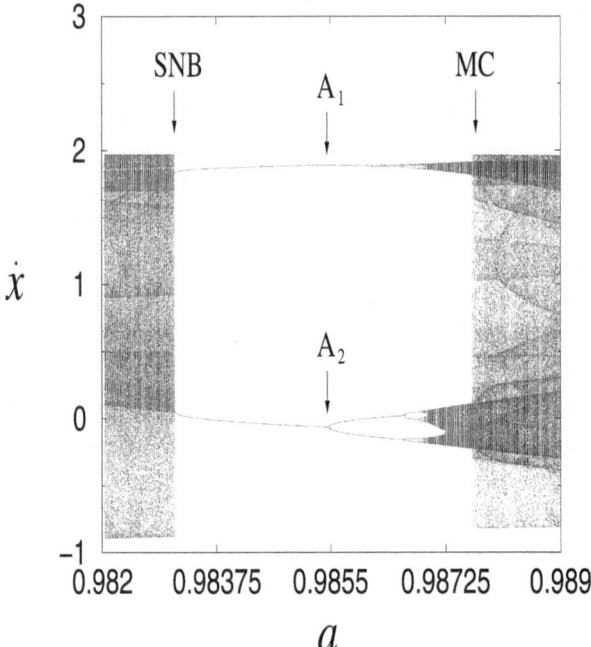

Fig. 4.1. Bifurcation diagram of $\dot{x}$ as a function of the driver amplitude a for attractors A_1 and A_2. MC denotes attractor merging crisis and SNB denotes saddle-node bifurcation. $\mu = 1$ and $\omega = 0.45$.

example, fixed points can be created or destroyed, or their stability can change. These changes in the system dynamics can be represented by the bifurcation diagram. A periodic window of the bifurcation diagram determined from the numerical solutions of equation (4.1) is shown in figure 4.1, where we plot $\dot{x}$ as a function of the driver amplitude a while keeping other control parameters fixed ($\mu = 1$ and $\omega = 0.45$). Within the periodic window, two (or more) attractors A_1 and A_2 co-exist, each with its own basin of attraction (Chian et al. 2005). At $a = 0.98312$, a period-1 limit cycle for each attractor A_1/A_2 is generated via a local saddle-node bifurcation (SNB), which evolves into a small chaotic attractor via a cascade of period-doubling bifurcations.

An attractor merging crisis occurs at the crisis point (MC), near $a = a_{MC} = 0.98765$. The phase-space trajectories of two small chaotic attractors (CA_1 and CA_2) in the phase space (x, $\dot{x}$), near the crisis point, are shown in figures 4.2(a) and 4.2(b), respectively. Note that CA_1 and CA_2 are symmetric with respect to each other. In fact, the

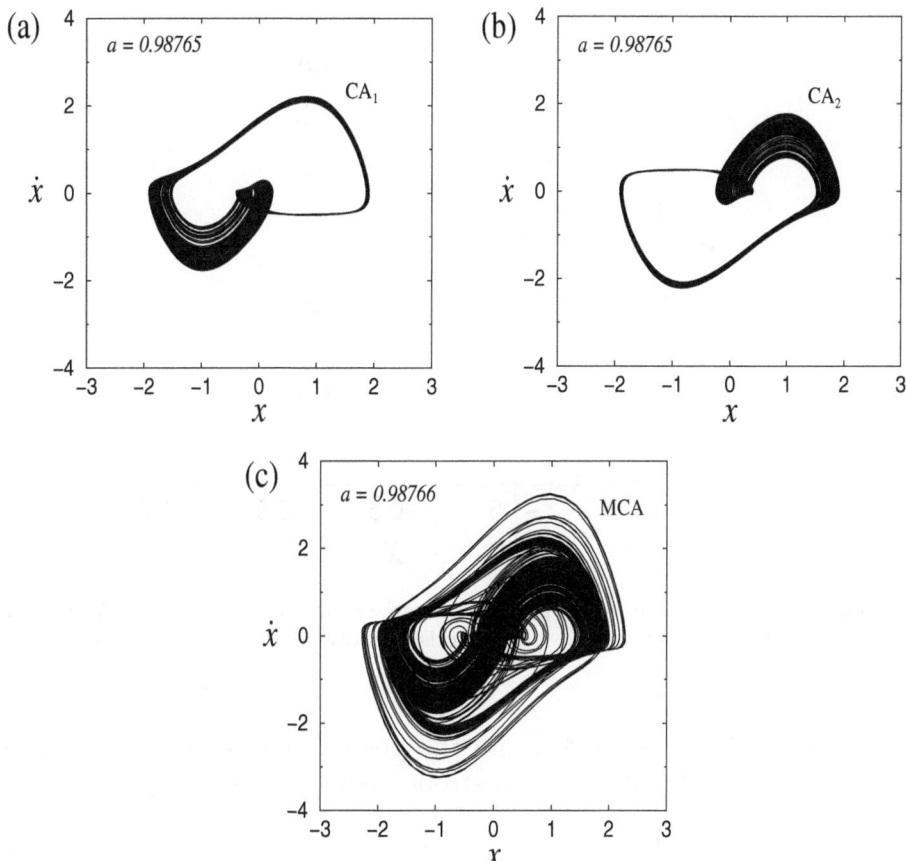

Fig. 4.2. Phase-space trajectories of: (a) pre-crisis chaotic attractor (CA$_1$) for $a =$ 0.98765, (b) pre-crisis chaotic attractor (CA$_2$) for $a = 0.98765$, (c) post-crisis merged chaotic attractor (MCA) for $a = 0.98766$

dynamic properties of these two co-existing attractors are identical. At the crisis point, each of the two small chaotic attractors simultaneously collide head-on with a period-3 mediating unstable periodic orbit on the boundary which separates their basins of attraction, leading to an attractor merging crisis due to a global bifurcation (Chian et al. 2005). As a consequence, the two pre-crisis small chaotic attractors merge to form a post-crisis large merged chaotic attractor (MCA), as seen in figure 4.2(c).

A Poincaré map of the phase-space trajectories of figure 4.2 is plotted in figure 4.3, which is a superposition of two pre-crisis weak chaotic

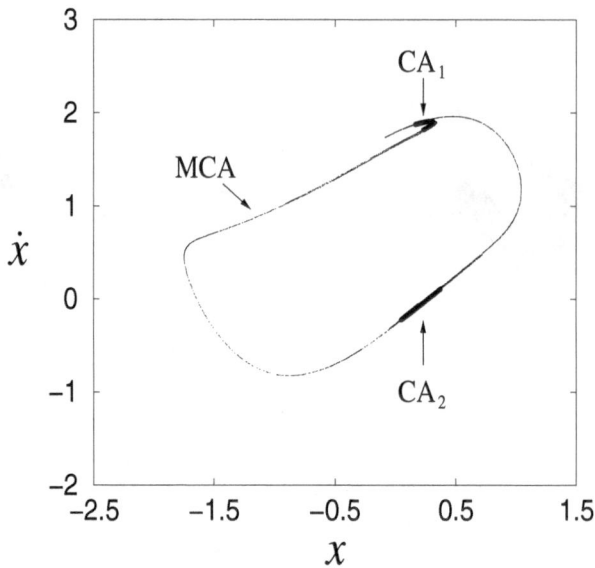

Fig. 4.3. Poincaré map of the post-crisis merged chaotic attractor (MCA, light line) for $a = 0.98766$, superposed by the pre-crisis chaotic attractors (CA$_1$ and CA$_2$, dark lines) for $a = 0.98765$

attractors (CA$_1$ and CA$_2$) and the post-crisis strong merged chaotic attractor (MCA). We define a stroboscopic Poincaré map

$$P : [x(t), \dot{x}(t)] \rightarrow [x(t+T), \dot{x}(t+T)], \qquad (4.2)$$

where $T = 2\pi/\omega$ is the driver period. Note that the two pre-crisis CA$_1$ and CA$_2$ are located in two small regions within the post-crisis MCA.

The time series of $\dot{x}$ for the two small chaotic attractors CA$_1$ and CA$_2$ at crisis, $a = 0.98765$, are shown in figures 4.4(a) and 4.4(b), respectively. The same time series of figures 4.4(a) and 4.4(b) plotted as a function of driver cycles are shown in figure 4.4(c). From figure 4.4(c), we see that before crisis the fluctuations of economic variables are weakly chaotic (laminar), localized in a small range of $\dot{x}$ (near $\dot{x} \sim 2$ and $\dot{x} \sim 0$), consistent with the Poincaré map in figure 4.3.

After the attractor merging crisis, there is only one large chaotic attractor (MCA) in the system. The time series of $\dot{x}$ of MCA after the crisis, for $a = 0.98766$ and $a = 0.988$, are shown in figures 4.5(a) and 4.5(b), respectively. The same time series plotted as a function of driver cycles are shown in figures 4.5(c) and 4.5(d), respectively. The time series in figure 4.5 show that the system dynamics becomes

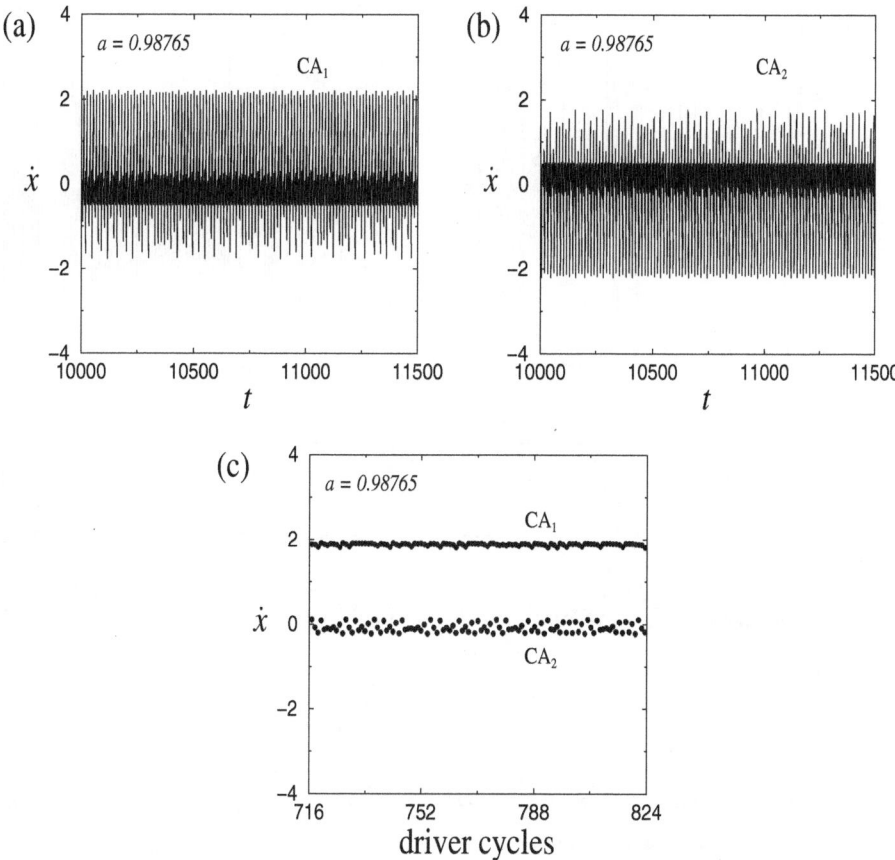

Fig. 4.4. Pre-crisis time series of $\dot{x}$ for $a = 0.98765$: (a) $\dot{x}(t)$ for chaotic attractor CA_1; (b) $\dot{x}(t)$ for chaotic attractor CA_2; (c) $\dot{x}$ as a function of driver cycles for (a) and (b), respectively

intermittent after the onset of attractor merging crisis, with periods of weakly chaotic (laminar) fluctuations interrupted abruptly by periods of strongly chaotic (bursty) fluctuations. A comparison of the time series of figures 4.4 and 4.5 indicates that the laminar phases in figure 4.5 are related to the pre-crisis attractors CA_1 and CA_2. Hence, the post-crisis system keeps memory of its weakly chaotic dynamics prior to crisis, and switches back and forth between the low-level fluctuations related to CA_1 and CA_2, linked by high-level fluctuations related to MCA. An examination of figure 4.5 shows that, as the system moves away from the crisis point, the average duration of laminar phases decreases and the regime switching becomes more frequent.

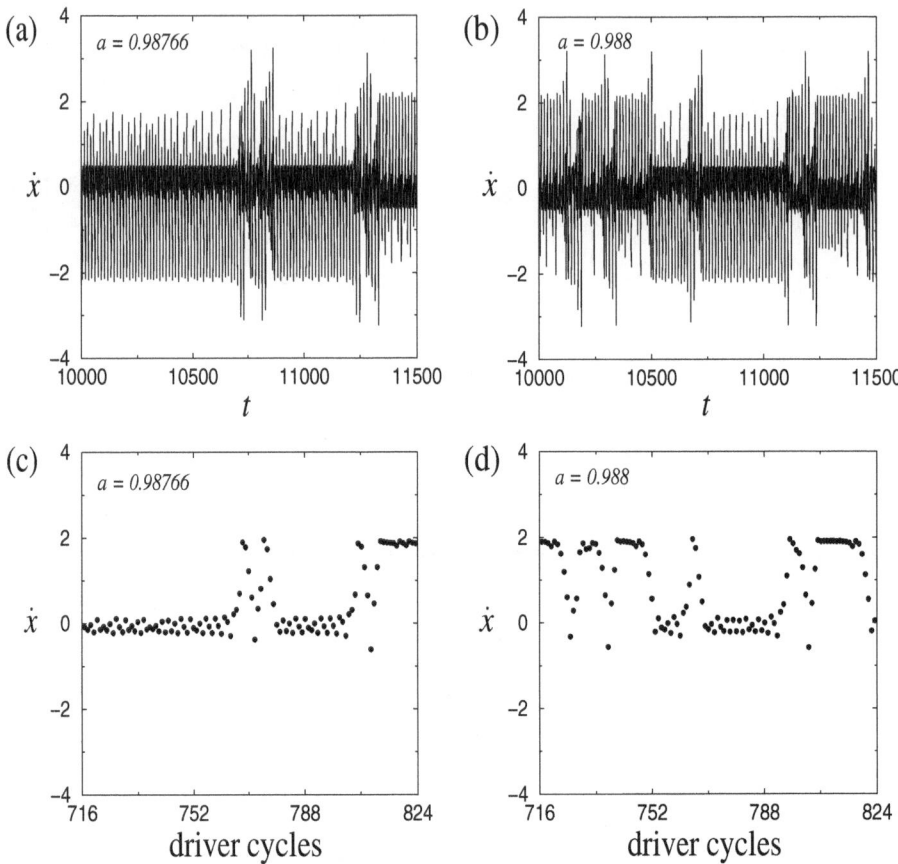

Fig. 4.5. Post-crisis intermittent time series of $\dot{x}$ for $a = 0.98766$ and $a = 0.988$. (a) and (b): $\dot{x}(t)$; (c) and (d): $\dot{x}$ as a function of driver cycles for (a) and (b), respectively.

The power spectra associated with the time series of figures 4.4 and 4.5 are shown in figure 4.6. It is evident that in all three cases the high-frequency portions of the spectra present power-law behaviors, which are typical features of intermittent financial systems such as stock markets and foreign exchange markets (Mantegna and Stanley 2000). A closer look of figures 4.6(a)-(c) shows that as the system becomes more chaotic, the discrete spikes of the power spectrum become less evident due to increasing multi-scale information transfer in the system.

The characteristic intermittency time, namely, the average duration of the laminar phases in the intermittent time series, depends on the value of the control parameter a. In the vicinity of the crisis point a_{MC} the average time spent by a path in the neighborhood of pre-crisis CA_1

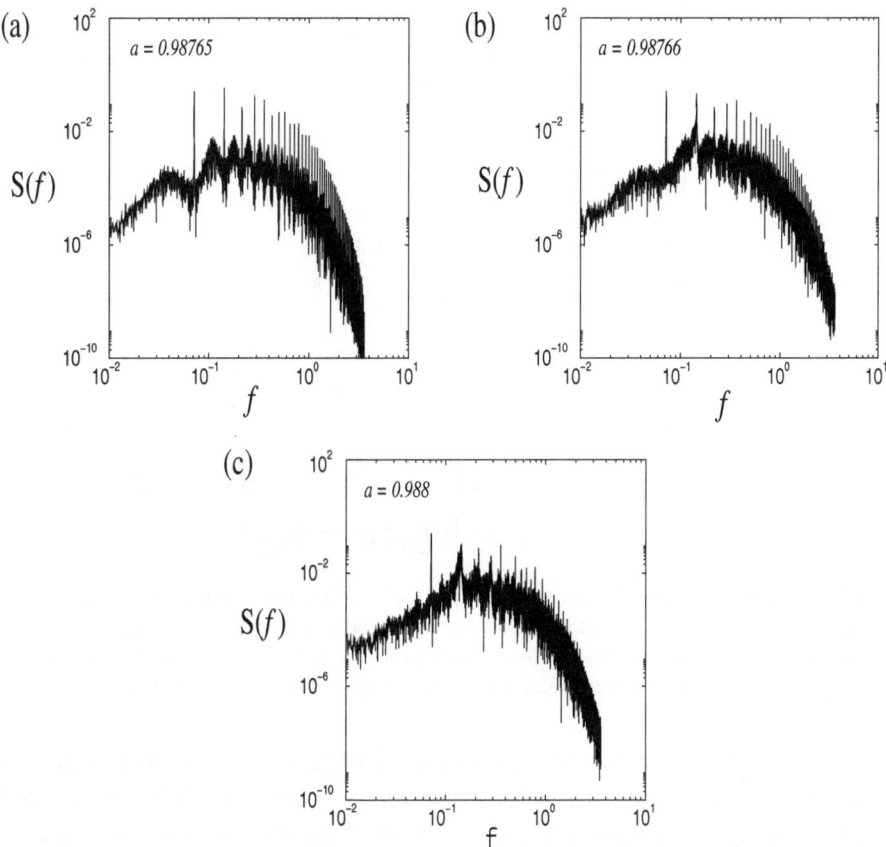

Fig. 4.6. Power spectrum $S(f)$ as a function of frequency f for: (a) $a = 0.98765$, (b) $a = 0.98766$, (c) $a = 0.988$

and CA_2 is very long (implying long memory), which decreases as a moves away from a_{MC} (implying shorter momory). The characteristic intermittency time (denoted by τ) can be calculated by averaging the duration of laminar phases related to CA_1/CA_2 over a long time series. Figure 4.7 is a plot of $\log_{10}\tau$ versus $\log_{10}(a - a_{MC})$, where the solid line with a slope $\gamma = -0.66$ is a linear fit. The squares (circles) denote the computed average time of the laminar phases related to CA_1 (CA_2). Note that circles and squares coincide most of the time, as expected from the symmetry of CA_1 and CA_2. Figure 4.7 reveals that the characteristic intermittency time τ decreases with the distance from the critical parameter, obeying a power-law scaling:

$$\tau \sim (a - a_{MC})^{-0.66}. \tag{4.3}$$

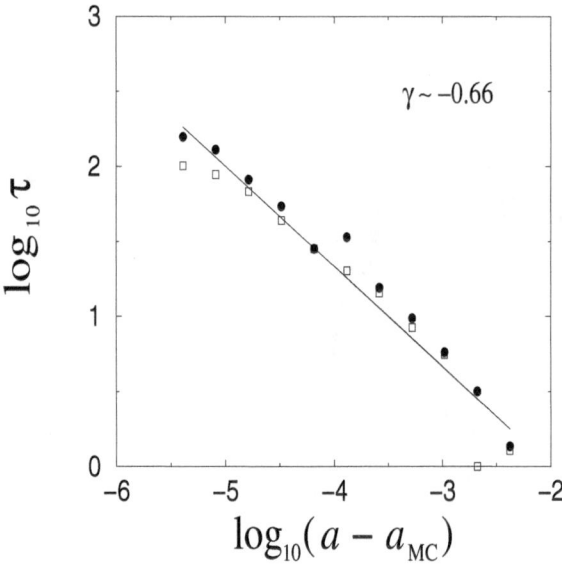

Fig. 4.7. Characteristic intermittency time as a function of the departure from the crisis point, $\log_{10} \tau$ versus $\log_{10}(a - a_{MC})$. The squares (circles) denote the computed average switching time from the laminar phases related to CA_1 (CA_2) to the bursty phases. The solid line is a linear fit of the computed values with a slope $\gamma = -0.66$.

The scaling relation for the van der Pol model of the economic type-I intermittency yields a scaling exponent of -0.074 (Chian et al. 2006). Comparing with equation (4.3), we see that the decrease of τ with the distance from the critical parameter for the economic crisis-induced intermittency is much faster than the economic type-I intermittency.

4.4 Concluding Comments

Forecasting the evolution of the complex system dynamics is the ultimate goal in economics. Chaos and nonlinear methods provide powerful tools to achieve this goal. For example, Bajo-Rubio et al. (1992) detected a chaotic behavior on daily time series of the Spanish Peseta-U.S. dollar exchange rate which allows short-run predictions. Soofi and Cao (1999) performed out-of-sample predictions on daily Peseta-U.S dollar spot exchange rates using a nonlinear deterministic technique of local linear predictor. Bordignon and Lisi (2001) proposed a method to evaluate the prediction accuracy of chaotic time series by means of prediction intervals and showed its effectiveness with data generated by a chaotic economic model.

A nonlinear prediction method being developed in population dynamics, weather dynamics and earthquake dynamics is based on attractor reconstruction in phase space using the time series of observed data (Drepper et al. 1994; Perez-Munuzuri and Gelpi 2000; Konstantinou and Lin 2004). This technique may be applied to economic forecasting. Information obtained from modeling intermittency of a complex economic system can guide the analysis of the reconstructed attractor by providing identifiable and predictable recurrent system patterns (Belaire-Franch 2004), and allowing the calculation of the characteristic intermittency time for each recurrent pattern. In particular, the determination of intermittent features in the modeled economic chaotic attractors, aided by the recognition of regions of high predictability in the chaotic attractors (Ziehmann et al. 2000), and the calculation of the power-law scaling in the intermittent error dynamics (Chu et al. 2002) may reduce prediction error and improve economic forecasting precision.

Economic forecasting relies on the agent's skill to recognize the patterns of recurrence in the past economic time series and to estimate the waiting time between bursts. Recurrence of unstable periodic structures is a manifestation of the memory dynamics of complex economic systems. Dynamical systems approach provides effective tools to identify the origin and nature of the recurrent patterns. In this chapter, we demonstrated how economic intermittency is induced by an attractor merging crisis and how to recognize different recurrent patterns in the intermittent time series of economic cycles by separating them into laminar (weakly chaotic) and bursty (strongly chaotic) phases. The characteristic intermittency time given by the scaling relation, equation (4.3), can be used to predict the turning points of regime switching from contrationary phases to expansionary phases in economic cycles.

Modeling of nonlinear economic dynamics enables us to obtain an in-depth knowledge of the nature of regime switiching and memory, in particular, their relation with each other. Econometric literatures on regime switching (Kirikos 2000; Bautista 2003; Kholodilin 2003) and long memory (Granger and Ding 1996; Resende and Teixeira 2002; Gil-Alana 2004; Muckley 2004) have evolved largely independently, as the two phenomena appear distint. Diebold and Inoue (2001) argued that regime switching and long memory are intimately related, which is in fact confirmed by our analysis. As an economic system evolves, microeconomic and macroeconomic instabilities lead to a variety of global and local bifurcations which in turn give rise to chaotic behaviors such as crisis-induced and type-I intermittencies. The techniques developed in this chapter can be applied to investigate intermittency in more complex economic models and to analyze other types of economic intermittency such as intermittency driven by a boundary crisis or an interior crisis, on-off intermittency, and noise-induced intermittency.

5

Attractor Merging Crisis in Nonlinear Economic Cycles

In this chapter, a numerical study is performed on a forced-oscillator model of nonlinear business cycles. In particular, an attractor merging crisis due to a global bifurcation is analyzed using the unstable periodic orbits and their associated stable and unstable manifolds. Characterization of crisis can improve our ability to forecast sudden major changes in economic systems.

5.1 Introduction

In recent years there is strong interest in the study of complex economic dynamics such as chaotic business cycles (Gabisch and Lorenz 1987; Puu 1989; Goodwin 1990; Lorenz 1993; Gandolfo 1997). Business cycles are fluctuations of macroeconomic variables resulting from instabilities in economic systems. Nonlinear evolution of economic instabilities leads to large-amplitude fluctuations of business cycles due to trajectories far-from-equilibrium. Complex systems approach provides a powerful tool to monitor and forecast the nonlinear dynamics of business cycles. For example, Mosekilde et al. (1992) studied the nonlinear mode-interaction between long-term and short-term business cycles; in a model of the economic long wave (Kondratiev cycle) driven by a periodic external forcing representing short-term business cycles, they identified nonlinear phenomena such as mode-locking, co-existence of attractors, period-doubling route to chaos, intermittent route to chaos, and crisis. Szydlowski, Krawiec and Tobola (2001) analyzed nonlinear oscillations in the Kaldor-Kalecki model of business cycles with

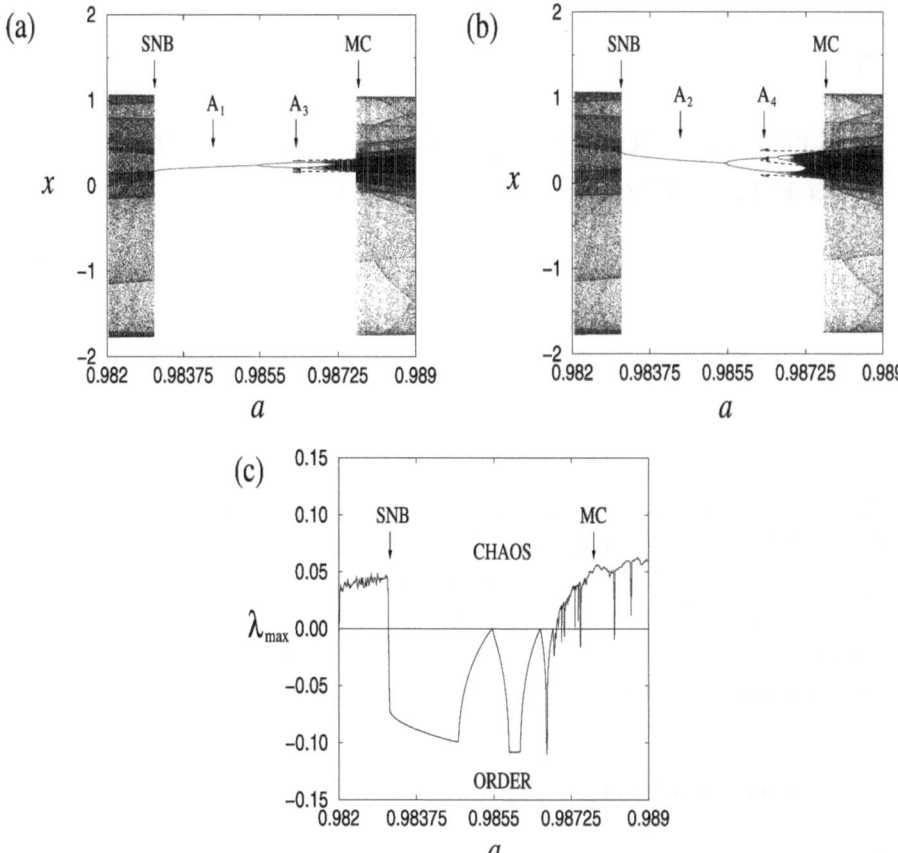

Fig. 5.1. Bifurcation diagram of x as a function of a for: (a) attractors A_1 and A_3, (b) attractors A_2 and A_4. (c) The maximum Lyapunov exponent λ_{max} as a function of a for either A_1 or A_2. MC denotes merging crisis; SNB denotes saddle-node bifurcation; the dashed lines denote the mediating unstable periodic orbits of period-3; $\omega = 0.45$, $\mu = 1$.

time lags in terms of bifurcation theory, and confirmed the existence of asymmetric cycles. Puu and Sushko (2004) employed a multiplier-accelerator model of business cycles, including a cubic nonlinearity, to study a number of bifurcation sequences for attractors and their basins of attraction.

Crisis is a global bifurcation resulting from the collision of a chaotic attractor with a mediating unstable periodic orbit or its associated stable manifold (Grebogi, Ott and York 1983; Grebogi et al. 1987; Chian, Borotto and Rempel 2002; Chian et al. 2002; Borotto, Chian and

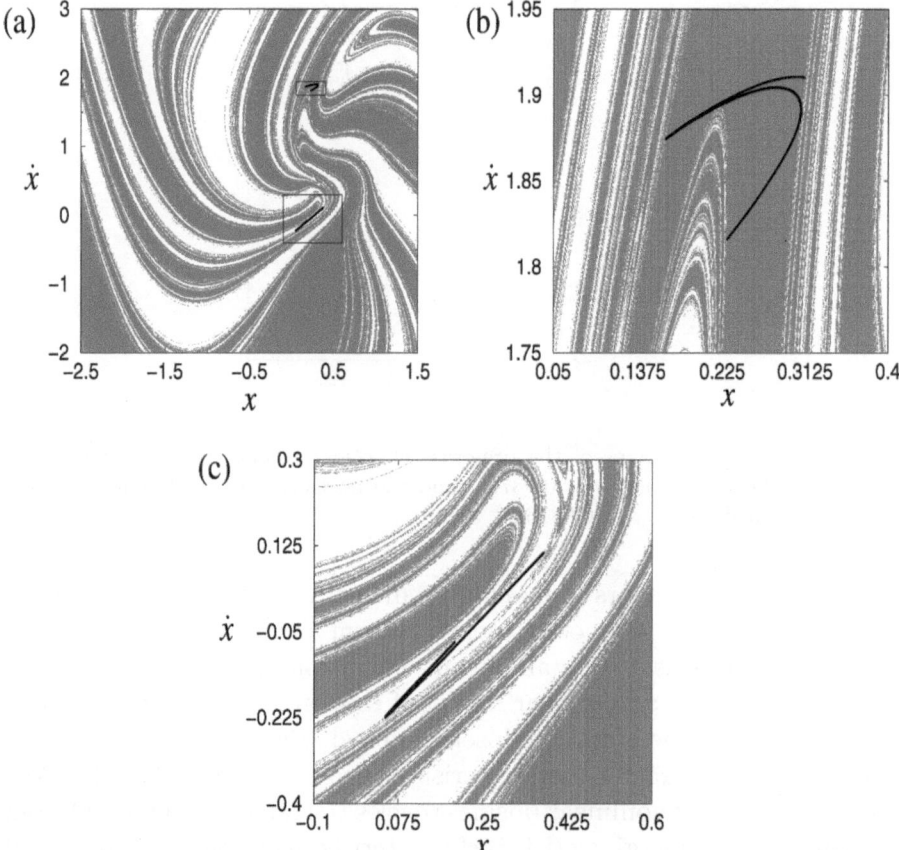

Fig. 5.2. (a) Basins of attraction at the crisis point $a = 0.98765$ for two co-existing attractors A_1 and A_2; (b) and (c) are the enlargements of the rectangular regions marked in (a); the gray regions denote the basins of attraction of A_1, the white regions denote the basins of attraction of A_2

Rempel 2004; Borotto et al. 2004) There are three types crises: boundary crisis, interior crisis and attractor merging crisis. A boundary crisis leads to a sudden appearance/disappearance of a chaotic attractor along with its basin of attraction, which occurs when the mediating unstable periodic orbit lies on the boundary between the basins of attraction of two attractors. An interior crisis leads to a sudden expansion/contraction of the chaotic attractor, when the collision between the chaotic attractor and the mediating unstable periodic orbit takes place in the interior of the basin of attraction of the attractor. An attractor

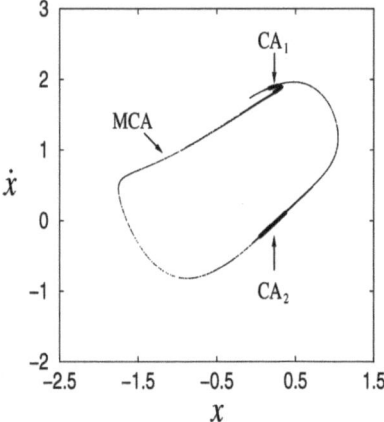

Fig. 5.3. Poincaré maps of the pre-crisis chaotic attractors (CA$_1$ and CA$_2$, dark lines) at the crisis point $a = 0.98765$, and the post-crisis merged chaotic attractor (MCA, light lines) at $a = 0.98766$

merging crisis appears in many systems with symmetries, whereby two (or more) chaotic attractors merge to form a single chaotic attractor.

An interior crisis, with an abrupt expansion of the chaotic attractor, was identified in a nonlinear model of economic long wave forced by a short-term business cycle (Mosekilde et al. 1992). In this chapter, we show that an attractor merging crisis appears in a forced van der Pol oscillator model of nonlinear business cycles (Chian et al. 2005). The onset of an attractor merging crisis is characterized using the tools of unstable periodic orbits and their associated stable and unstable manifolds.

5.2 Nonlinear Model of Economic Cycles

We adopt the driven van der Pol (VDP) differential equation to model the nonlinear dynamics of business cycles under the action of a periodic exogenous force

$$\ddot{x} + \mu(x^2 - 1)\dot{x} + x = a\sin(\omega t). \tag{5.1}$$

The equilibrium solution of the VDP equation reduces to a repeller fixed point located at the origin $(0, 0)$ in the phase space $(x, dx/dt)$. In the absence of exogenous forcing ($a = 0$), the asymptotic solution of equation (5.1) is a limit cycle surrounding the equilibrium fixed point. In the presence of exogenous forcing, either periodic (orderly) or aperiodic (chaotic) solutions appear when we vary any of three control parameters:

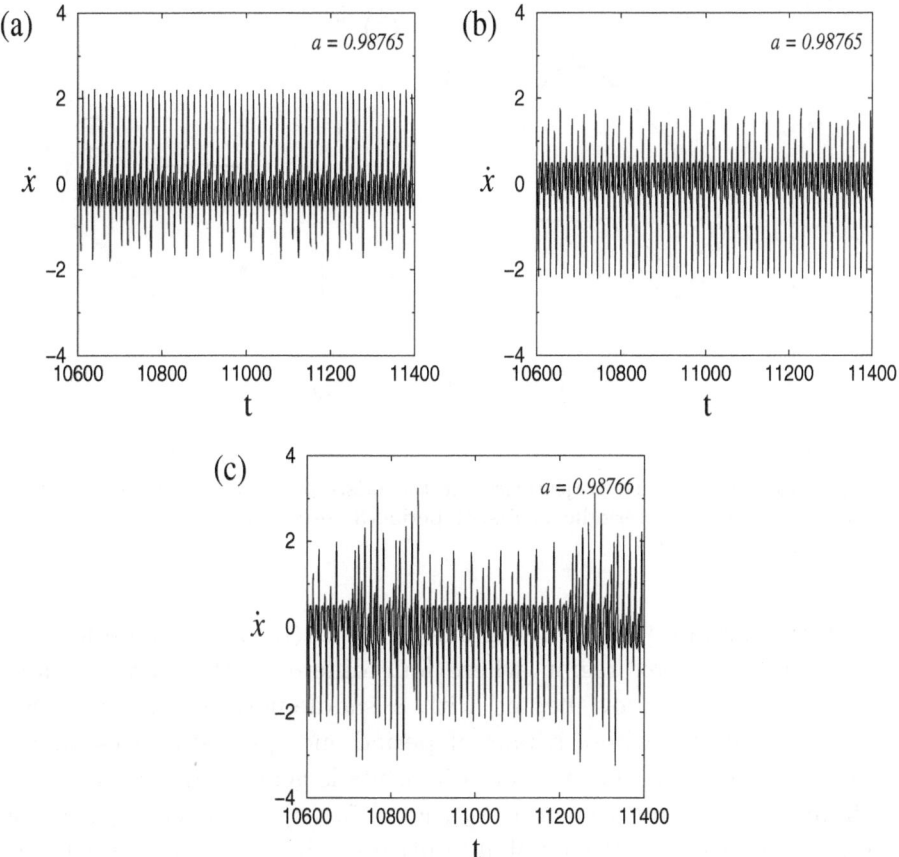

Fig. 5.4. Time series $\dot{x}(t)$ for: (a) chaotic attractor CA_1 at $a = 0.98765$, (b) chaotic attractor CA_2 at $a = 0.98765$, (c) merged chaotic attractor MCA at $a = 0.98766$

a, ω, μ. The VDP equation (5.1) (when $a = 0$) is invariant under the flip operation $(x \rightarrow -x)$. This symmetry is a typical property of dynamical systems that exhibit attractor merging crises (Grebogi et al. 1987).

5.3 Attractor Merging Crisis

In order to obtain a global view of the system dynamics, we construct a bifurcation diagram from the numerical solutions of equation (5.1) by varying the control parameter a while keeping the other two control parameters fixed ($\mu = 1, \omega = 0.45$). The Poincaré plane is defined by

$$P : x(t) \rightarrow x(t + T), \qquad (5.2)$$

where $T = 2\pi/\omega$ is the driver period.

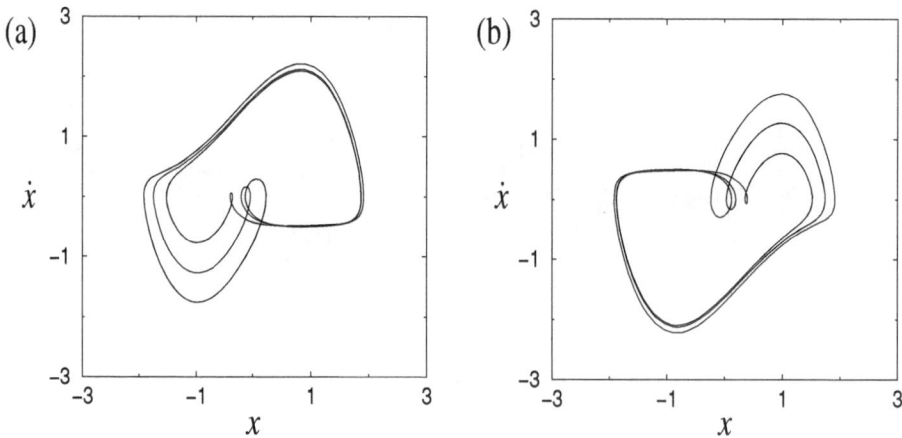

Fig. 5.5. Phase space trajectories, at the crisis point $a = 0.98765$, of the two mediating unstable periodic orbits of period-3 responsible for attractor merging crisis of: (a) A_1, (b) A_2

The bifurcation diagrams figures 5.1(a)-5.1(b) display a periodic window in a complex region where four different attractors are found. The periodic window begins with a saddle-node bifurcation (SNB) at $a = 0.98312$, where a pair of period-one (p-1) stable (solid lines) and unstable (not shown) periodic orbits is generated for attractor A_1 (figure 5.1(a)) and attractor A_2 (figure 5.1(b)), respectively; the periodic window ends with a global bifurcation due to an attractor merging crisis (MC) at $a_{MC} = 0.98765$, where the two chaotic attractors CA_1 and CA_2 combine to form a merged chaotic attractor (MCA). The rich dynamical states displayed by the bifurcation diagram indicate that a dynamical system is sensitively dependent on a small variation of its control parameters.

As we increase a after the saddle-node bifurcation (SNB), the stable periodic orbit (SPO) of A_1 (A_2) undergoes a cascade of period-doubling bifurcations leading to a chaotic attractor CA_1 (CA_2). Figure 5.1(a) (5.1(b)) shows that a second attractor A_3 (A_4) coexists with A_1 (A_2), respectively, for a small range of the control parameter, between $a = 0.9862400$ and $a = 0.9864085$. Attractor A_3 (A_4) is created by a saddle-node bifurcation, where a pair of p-3 stable (solid lines) and unstable (dashed lines) periodic orbits is generated. A_3 and A_4 are destroyed at $a = 0.9864085$ due to a boundary crisis (see e.g., Chian, Borotto and Rempel 2002).

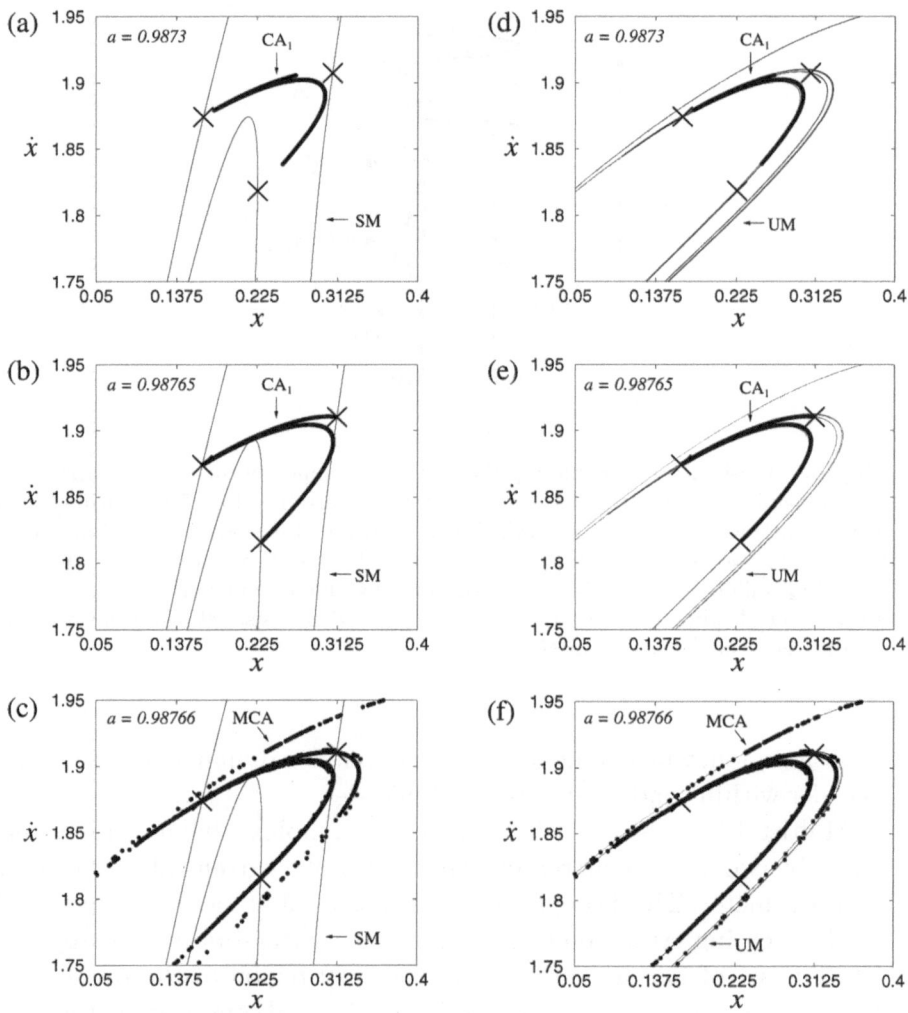

Fig. 5.6. Poincaré map in the vicinity of CA$_1$ (same region as Fig. 5.2(b)). (a) and (d): before crisis ($a = 0.9873$), (b) and (e): at crisis ($a = 0.98765$), and (c) and (f): after crisis ($a = 0.98766$). The crosses denote the Poincaré points of the mediating unstable periodic orbit of period-3; the dark lines (dark points) denote the chaotic attractors (CA$_1$ and MCA); the light lines denote the stable/unstable manifolds (SM/UM) of the mediating saddle.

The corresponding behavior of the maximum Lyapunov exponent for either A$_1$ or A$_2$, calculated by the Wolf algorithm (Wolf 1985), is shown in figure 5.1(c). Figure 5.1 shows that there are many chaotic regions within a periodic window and there are many periodic windows within a

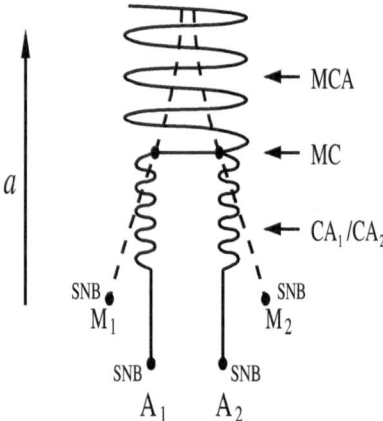

Fig. 5.7. Crisis diagram depicting the system dynamics as the control parameter a varies. Saddle-node bifurcations (SNB) occur at a certain value of a, creating two coexisting attractors (A_1 and A_2), which via a cascade of period-doubling bifurcations evolve into two chaotic attractors (CA_1 and CA_2). At the crisis point (MC), CA_1 and CA_2 collide head-on with the mediating unstable periodic orbits (M_1 and M_2), respectively, leading to the onset of attracting merging crisis (MC) and the formation of a merged chaotic attractor (MCA).

chaotic region, which indicates that in a complex dynamical system there is order within chaos and chaos within order.

Multistability is a basic feature of complex dynamical systems whereby two or more attractors can coexist for a given value of the control parameter. This is depicted by the basins of attraction in figure 5.2, at the merging crisis point MC, where two attractors A_1 and A_2 coexist. The set of initial conditions in the gray region of the phase space $(x, dx/dt)$ will lead to A_1, whereas the set of initial conditions in the white region will lead to A_2, as clarified in the enlarged plots, figures 5.2(b) and 5.2(c), respectively, of the two rectangular regions marked in figure 5.2(a).

After the attractor merging crisis, the two pre-crisis chaotic attractors (CA_1 and CA_2) of figure 5.2(a) combine to form a merged chaotic attractor (MCA), as shown in figure 5.3 in the Poincaré plane. The merged attractor after crisis is larger than the union of the two attractors before crisis. Time series plots of the economic variable $\dot{x}(t)$ at pre-crisis are given in figure 5.4(a) for CA_1 and figure 5.4(b) for CA_2, respectively, and at post-crisis (MCA) is given in figure 5.4(c). The amplitudes of business cycle fluctuations after the attracting merging crisis are much larger than before the crisis.

Unstable periodic orbit (UPO) plays a key role at the onset of attractor merging crisis. We numerically determine UPO from the numerical solution of equation (5.1) using the Newton algorithm. Analysis shows that the mediating p-3 unstable periodic orbits (M), evolved from the saddle-node bifurcations at the birth of A_3 (A_4), are responsible for the attractor merging crisis. The dashed lines in figures 5.1(a)-5.1(b) denote M. The phase space trajectory of the two mediating UPOs that collide with A_1 (A_2), respectively, at the crisis point MC are displayed in figure 5.5. Note that the two UPOs in figure 5.5 are symmetric under reflection off x- and y- axis. This is a manifestation of the symmetry property of the VDP equation (5.1).

Characterization of crisis in economic dynamics can be performed using the Poincaré method. On the Poincaré plane, an UPO transforms into a saddle fixed point with its associated stable and unstable manifolds. Figure 5.6 displays the dynamical states of chaotic business cycles on the Poincaré plane in the vicinity of A_1 (same region as figure 5.2(b)) before (figure 5.6(a)), at (figure 5.6(b)), and after (figure 5.6(c)) the onset of attractor merging crisis, respectively. The crosses denote the three fixed points of the p-3 mediating saddle. The dark lines (and points) denote the chaotic attractor, and the light lines in figures 5.6(a)-5.6(c) denote the numerically computed stable manifolds of the mediating saddle. Evidently, figure 5.6(b) demonstrates the head-on collision, at the crisis point MC, of the chaotic attractor with the mediating saddle and its stable manifolds. This collision leads to the formation of a merged chaotic attractor, seen in figure 5.6(c).

Figures 5.6(d)-(f) displays the same system dynamics of figure 5.6(a)-(c), with the stable manifolds replaced by the numerically computed unstable manifolds (UM) of the mediating saddle. Our numerical calculations render support to the conjecture of Parker and Chua (Parker 1989) and Ott (Ott 1993) that a chaotic attractor contains the unstable manifolds of every UPO of the chaotic attractor. Figure 5.6(f) demonstrates that the post-crisis chaotic attractor in fact coincides with the closure of the unstable manifolds of the mediating saddle. Although we only show the dynamical states of A_1 in figure 5.6, the same behavior also applies to A_2 due to the symmetry of the VDP system.

Examination of figure 5.2 shows that at the onset of attractor merging crisis, the attractors collide head-on with the boundary of the basins of attraction that separate attractors A_1 and A_2. This boundary coincides with the stable manifolds of the mediating saddle, as demonstrated by figures 5.2(b) and 5.6(b). Hence, figures 5.2 and 5.6 provide two alternative ways of characterizing the onset of attractor merging crisis.

5.4 Concluding Comments

This chapter shows that chaotic transitions such as the attractor merging crisis is a fundamental feature of nonlinear business cycles. The crisis diagram for the attractor merging crisis studied is given in figure 5.7, which summarizes the system dynamics leading to the onset of crisis. Mathematical modelling of crisis can deepen our understanding of sudden major changes of economic variables often encountered in business cycles. The techniques developed in this chapter for crisis characterization (e.g., figures 5.2 and 5.6) can contribute to improve the prediction of the onset of abrupt major changes in business cycles as well as other economic systems.

Attractor merging crisis appears in systems with symmetry such as equation (5.1). This type of crisis is absent when the system symmetry is broken. However, other types of crisis phenemena such as boundary crisis (Chian, Borotto and Rempel 2002) and interior crisis (Borotto, Chian and Rempel 2004) can be found in asymmetric systems such as the asymmetric van der Pol equation (Engelbrecht and Kongas 1995), and are in fact present in the solutions of equation (5.1). The techniques developed in this chapter can be readily applied to characterize boundary and interior crises. Hence, crises and global bifurcations are ubiquitous in either symmetric or asymmetric nonlinear economic systems.

6

Chaotic Transients in Nonlinear Economic Cycles

6.1 Introduction

In chapter 2, we showed that a nonlinear economic system is intrinsically unstable; as the endogenous or exogenous parameters are varied, the system undergoes a variety of local and global bifurcations such as saddle-node bifurcation and attractor merging crisis, seen in the periodic window in figure 2.5. Chapter 3 showed that saddle-node bifurcation is a route from order to chaos, leading to a chaotic dynamical behavior known as type-I intermittency. Chapter 5 analyzed an attractor merging crisis in chaotic business cycles which leads to a transition from weak chaos to strong chaos; the strong chaos exhibits a dynamical behavior known as crisis-induced intermittency, as seen in chapter 4. In this chapter, we will study the roles of unstable periodic orbits and chaotic saddles in type-I intermittency and crisis-induced intermittency in complex economic systems (Chian, Rempel and Rogers 2006), based on the forced van der Pol oscillator model of nonlinear economic cycles

$$\ddot{x} + \mu(x^2 - 1)\dot{x} + x = a \sin(\omega t). \tag{6.1}$$

6.2 Chaotic Saddle

Chaotic sets are not necessarily attracting sets. A set of unstable periodic orbits can be chaotic and nonattracting so that the orbits in the neighborhood of this set are eventually repelled from it; nonetheless, this set can contain a chaotic orbit with at least one positive Lyapunov exponent (Nusse and York 1989). If the chaotic orbit has also one negative Lyapunov exponent the nonattracting set is known as a chaotic

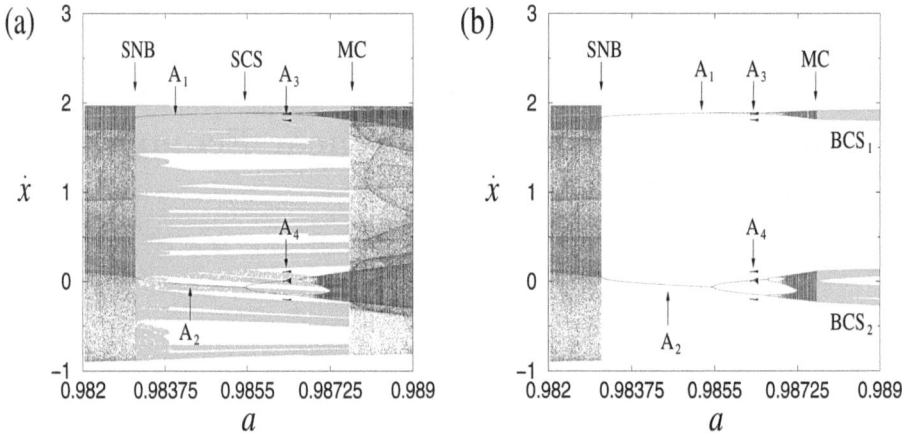

Fig. 6.1. Chaotic saddle in bifurcation diagram. (a) Bifurcation diagram, $\dot{x}$ as a function of a, for attractors (dark) A_0, A_1, A_2, A_3 and A_4, and the surrounding chaotic saddle SCS (gray); (b) bifurcation diagram of (a) and the band chaotic saddle BCS (gray). SNB denotes saddle-node bifurcation, MC denotes attractor merging crisis.

saddle. Both chaotic saddles and chaotic attractors are composed of unstable periodic orbits.

Figure 6.1(a) shows a bifurcation diagram for both attractors (dark) and chaotic saddles (gray) for a periodic window (same as figure 2.5(a)), where we plot $\dot{x}$ as a function of the amplitude a of the exogenous forcing while keeping other control parameters constant ($\mu = 1$ and $\omega = 0.45$). As seen in chapter 2, within this periodic window, two or more attractors can coexist. To plot the chaotic saddle, for each value of the control parameter a, we plot a straddle trajectory close to the chaotic saddle using the PIM triple algorithm (Nusse and Yorke 1989, Rempel et al. 2004a,b). The periodic window in figure 6.1(a) begins with a saddle-node bifurcation (SNB) at $a_{SNB} = 0.98312$, where a pair of period-1 stable and unstable periodic orbits for each attractor (A_1 and A_2) is created, respectively. As we increase a, the pair of period-1 stable periodic orbits undergoes a cascade of periodic-doubling bifurcations which leads to the formation of a pair of weakly chaotic attractors localized in two separate bands in the bifurcation diagram. We call the region of the phase space occupied by the attractor throughout the periodic window the band region, and the region occupied by the chaotic saddle, the surrounding region (Szabó et al. 2000). Trajectories started in the surrounding region usually behave chaotically for a finite transient time while traversing in the vicinity of the surrounding chaotic saddle (SCS), after which they converge to the attractor. The transient time is related to the structure of SCS and its manifolds. Like a saddle point, chaotic saddles possess a stable and an unstable manifold. The stable manifold of a chaotic saddle is the sets of points that converge to the chaotic saddle in forward time, and the unstable manifold is the sets of points that converge to the chaotic saddle in the time reverse dynamics (Nusse and Yorke 1989). Initial conditions close to the stable manifold are first attracted to SCS and stay close to its neighborhood for sometime, before they are repelled by its unstable manifold. The closer an initial condition is to the stable manifold, the longer its transient time. Note from the bifurcation diagram in figure 6.1(a) that as the control parameter a varies, the dynamics of the surrounding chaotic saddle also undergoes considerable changes.

The end of the periodic window in figure 6.1(a) is marked by an attractor merging crisis (MC) at $a_{MC} = 0.98765$, where the two banded weakly chaotic attractors merge to form a strongly chaotic attractor. Figure 6.1(a) shows that for a small range of the control parameter, between $a = 0.9862400$ and 0.9864085, attractors A_3 and A_4 coexist with A_1 and A_2. Attractors A_3 and A_4 are created by a saddle-node bifurcation at $a = 0.9862400$, where a pair of period-3 stable and unstable periodic orbits are generated. A_3 and A_4 are destroyed by a boundary crisis at $a = 0.9864085$. We will demonstrate later that the attractor merging crisis (MC) at $a = 0.98765$ arises from the collision of the two banded weakly chaotic attractors with the pair of period-3 mediating unstable periodic orbits created at $a = 0.9862400$. Right after the attractor merging crisis, the pair of weakly chaotic attractors lose their asymptotic stability and are converted into a pair of chaotic saddles in the band regions, as shown in figure 6.1(b). It is worth pointing out that, although in figure 6.1(a) we plot the surrounding chaotic saddle only inside the periodic window, it is actually present throughout the whole bifurcation diagram. In the chaotic regions beyond SNB and MC, the chaotic saddles are embedded in the chaotic attractor A_0.

6.3 Chaotic Transient

Figure 6.2(a) shows the Poincaré map of the surrounding chaotic saddle SCS (gray) obtained by the PIM triple algorithms (Nusse and Yorke 1989, Rempel et al. 2004a,b) in the beginning of the periodic window,

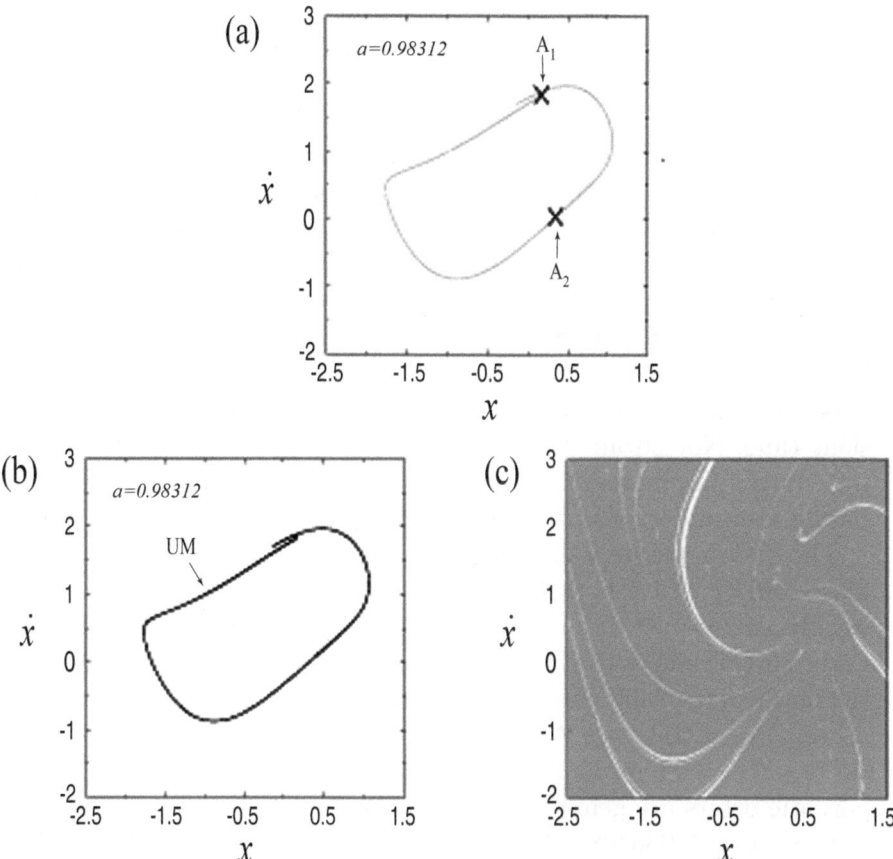

Fig. 6.2. Poincaré maps of chaotic saddle leading to periodic attractor. (a) Poincaré map of the surrounding chaotic saddle (gray) and the pair of period-1 fixed points (cross) A_1 and A_2 for $a = 0.98312$, (b) the unstable manifold (UM) of the surrounding chaotic saddle, (c) the stable manifold (gray) of the surrounding chaotic saddle.

superposed by the pair of period-1 periodic attractors A_1 and A_2 (cross) at $a_{SNB} = 0.98312$. Figures 6.2(b) and 6.2(c) display the unstable and stable manifolds, respectively, of the surrounding chaotic saddle of figure 6.2(a), found by the sprinkler algorithm (Kantz and Grassberger 1985, Hsu, Ott and Grebogi 1988, Rempel et al. 2004a,b). Figures 6.1(a) and 6.1(b) show that the chaotic saddles have gaps which reflect the fractal structure of a chaotic saddle along its unstable direction. The presence of gaps in the chaotic saddle can be seen in figure 6.2(a). It follows from figure 6.2 that a chaotic saddle is formed by the

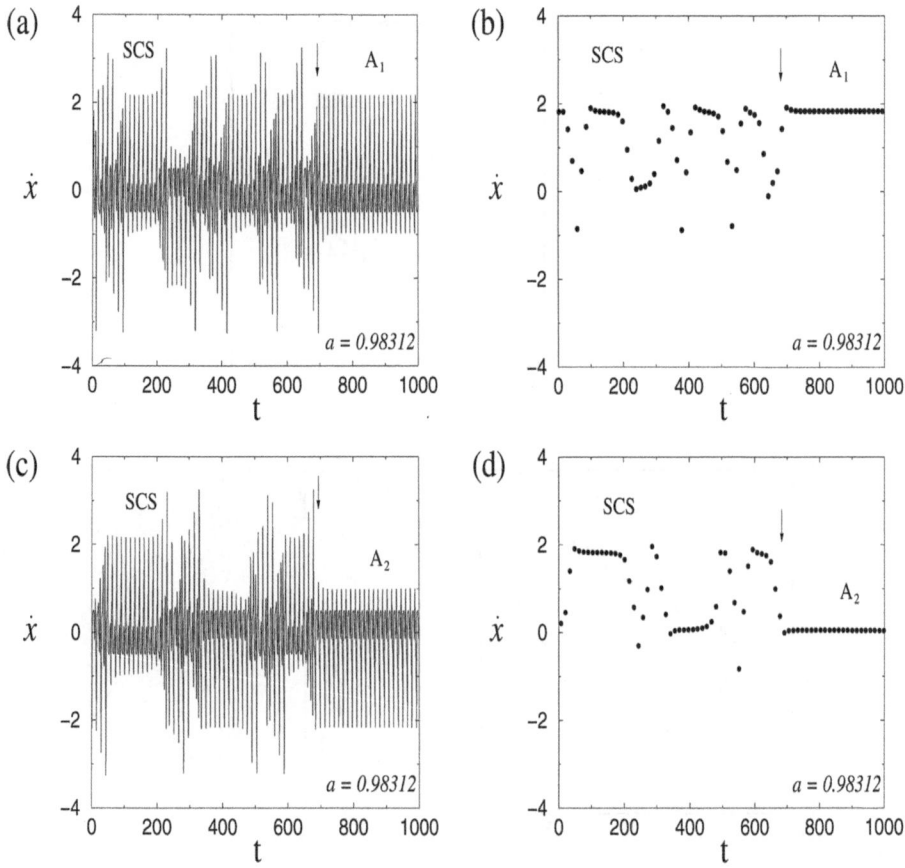

Fig. 6.3. Time series of chaotic transient leading to periodic attractor. (a) Time series, $\dot{x}$ as a function of t, of a chaotic transient (SCS) that converges to a periodic time series of period-1 attractor A_1 for $a = 0.98312$ after the time indicated by the arrow; (b) the same time series of (a) plotted as a function of the driver cycles; (c) time series of a chaotic transient (SCS) that converges to a periodic time series of period-1 attractor A_2 for $a = 0.98312$ after the time indicated by the arrow; (d) the same time series of (c) plotted as a function of the driver cycles.

intersection of its stable and unstable manifolds. The empty space between the intersection points along the unstable direction is the origin of the gaps in the chaotic saddle. Inside the periodic window the gaps of the chaotic saddle are empty in the sense that they do not contain unstable periodic orbits, only nonrecurrent points whose orbits converge very quickly to the small neighborhood of the period-1 attractors (Robert et al. 2000). Figure 6.3 shows examples of the time series of the

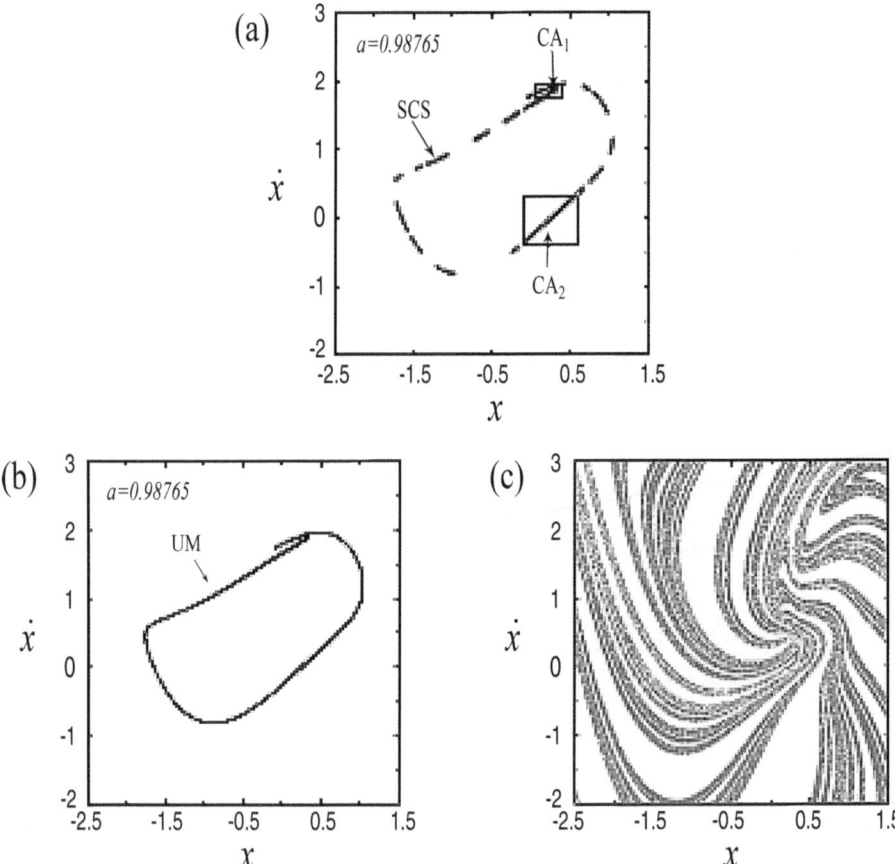

Fig. 6.4. Poincaré maps of chaotic saddle leading to chaotic attractor. (a) Poincaré maps of the surrounding chaotic saddle (gray) and the pair of weak chaotic attractors CA_1 and CA_2 for $a = 0.98765$, (b) the unstable manifold (UM) of the surrounding chaotic saddle, (c) the stable manifold (gray) of the surrounding chaotic saddle.

trajectory at $a_{SNB} = 0.98312$. For an arbitrary initial condition, the trajectory stays a finite transient period in the neighborhood of the surrounding chaotic saddle SCS until it converges to either of the period-1 periodic attractor A_1 (figures 6.3(a) and 6.3(b)) or A_2 (figures 6.3(c) and 6.3(d)) at the time indicated by the arrow, depending on the initial condition. Thus, inside the periodic window the surrounding chaotic saddle plays the role of chaotic transient motion before converging to the attractor.

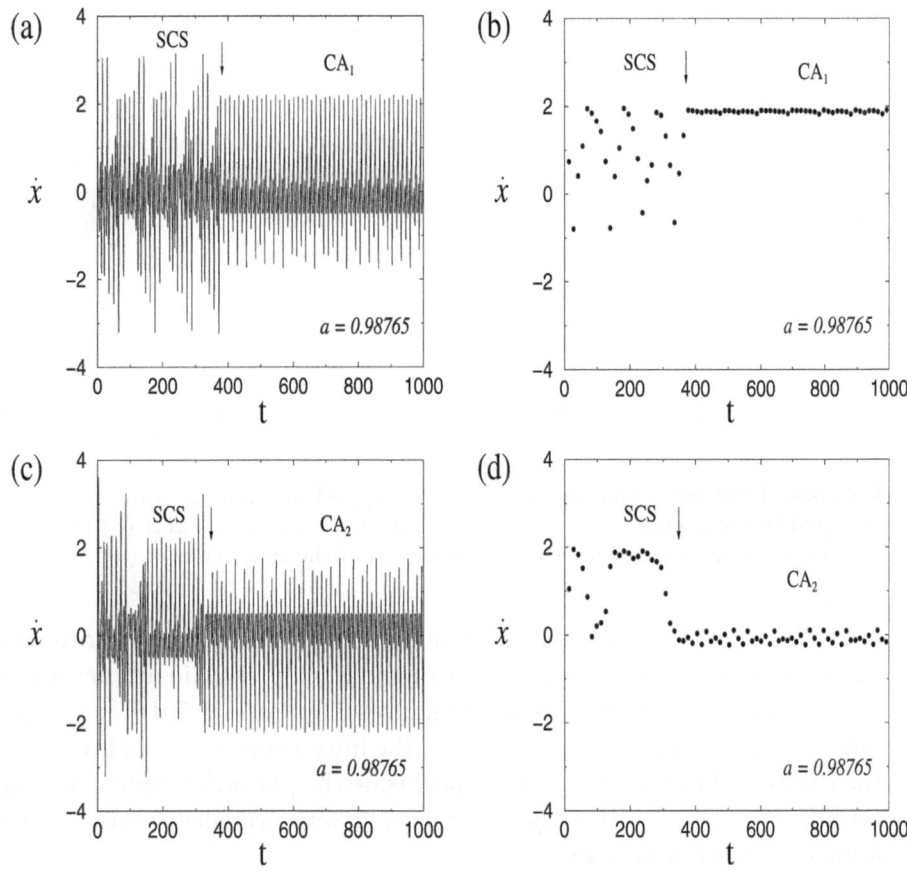

Fig. 6.5. Time series of chaotic transient leading to chaotic attractor. (a) Time series, $\dot{x}$ as a function of t, of a chaotic transient (SCS) that converges to a chaotic time series of the weak chaotic attractor CA_1 for $a = 0.98765$ after the time indicated by the arrow; (b) the same time series of (a) plotted as a function of the driver cycles; (c) time series of a chaotic transient (SCS) that converges to a chaotic time series of the weak attractor CA_2 for $a = 0.98765$ after the time indicated by the arrow; (d) the same time series of (c) plotted as a function of the driver cycles.

Figure 6.4(a) shows the Poincaré map of the surrounding chaotic saddle SCS (gray) obtained at the end of the periodic window, superposed by the pair of weakly chaotic attractors CA_1 and CA_2 (black) at $a_{MC} = 0.98765$. Figures 6.4(b) and 6.4(c) display the unstable and stable manifolds, respectively, of the surrounding chaotic saddle of figure 6.4(a). Figure 6.5 shows examples of the time series of the trajectory at $a_{MC} = 0.98765$. For an arbitrary initial condition, the trajectory stays a finite

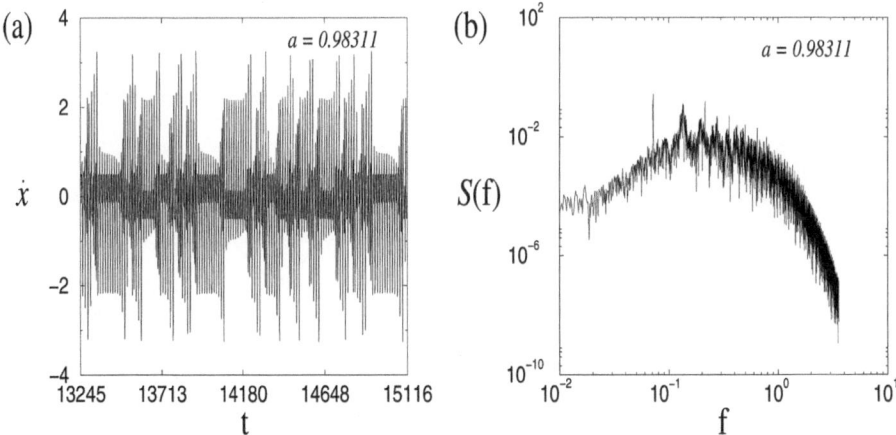

Fig. 6.6. Time series and power spectrum of type-I intermittency. (a) Time series of type-I intermittency, $\dot{x}$ as a function of time t, for $a = 0.98311$; (b) the power spectrum, $|\dot{x}|^2$ as a function of the frequency f, of the time series of (a).

transient period in the neighborhood of the surrounding chaotic saddle SCS until it converges to either of the weakly chaotic attractor CA_1 (figures 6.5(a) and 6.5(b)) or CA_2 (figures 6.5(c) and 6.5(d)) at the time indicated by the arrow, depending on the initial condition. This confirms the results of figures 6.2 and 6.3, that inside the periodic window, the surrounding chaotic saddle plays the role of chaotic transient motion before approaching an attractor.

6.4 Unstable Structures in Type-I Intermittency

Next let's turn our attention to the role of chaotic saddles in the chaotic regions of figure 6.1. As shown by chapter 3, the chaotic attractor prior to the onset of the saddle-node bifurcation, to the left of $a_{SNB} = 0.98312$ in figure 6.1, exhibits type-I intermittency whereby the time series of economic variables switch episodically back and forth between periods of apparently periodic and bursting chaotic fluctuations, exemplified in figure 6.6(a); the corresponding power spectrum has a power-law behavior at high-frequencies as shown in figure 6.6(b), typical of real intermittent financial data.

We saw in figure 6.2(a) that at the onset of saddle-node bifurcation at $a_{SNB} = 0.98312$ there is a surrounding chaotic saddle (SCS) which represents the chaotic transient preceding convergence to the period-1 periodic attractors A_1 and A_2. Note that there are

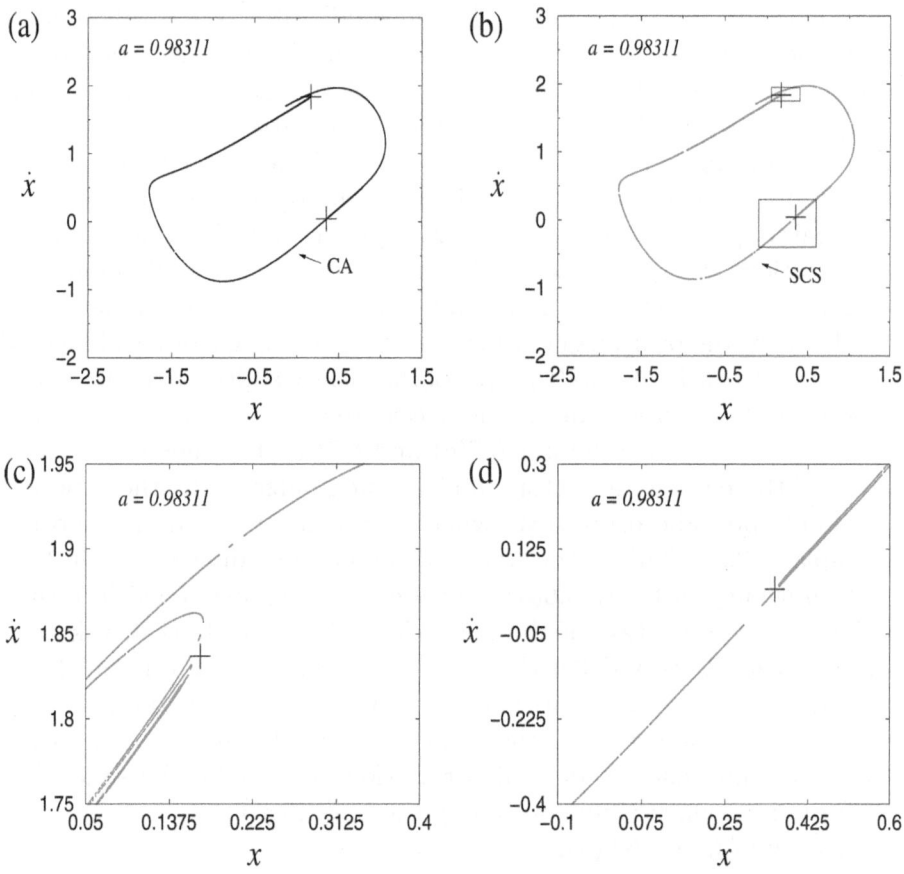

Fig. 6.7. Chaotic attractor and chaotic saddle in type-I intermittency. (a) Poincaré map of the chaotic attractor (CA) for type-I intermittency at $a = 0.98311$, (b) the surrounding chaotic saddle (SCS) embedded in the chaotic attractor of (a), (c) and (d) are enlargements of the two rectangular regions of (b). The cross denotes the pair of period-1 unstable periodic orbits created at the saddle-node bifurcation $a = 0.98312$.

gaps in the surrounding chaotic saddle in figure 6.2(a). As the system undergoes a transition from order to chaos via a saddle-node bifurcation, the surrounding chaotic saddle (SCS) is converted into a chaotic attractor (CA) as shown in the Poincaré map in figure 6.7(a) for $a = 0.98311$, where we also plotted the fixed points (cross) of the pair of period-1 unstable periodic orbits (M) created by the saddle-node bifurcation at $a_{SNB} = 0.98312$. Since the unstable periodic orbits are robust, all the unstable periodic orbits (with the exception of M) contained in the surrounding chaotic saddle after the saddle-node bifurcation (figure 6.2(a)) continue to exist in the chaotic region beyond the saddle-node bifurcation (to the left of a_{SNB}). Thus, the surrounding chaotic saddle is embedded in the chaotic attractor of figure 6.7(a), as shown in figure 6.7(b). An enlargement of the rectangular regions of figure 6.7(b) is given in figures 6.7(c) and 6.7(d), respectively. Although the pair of period-1 saddle points (M) appear only after the saddle-node bifurcation, the system keeps the memory of these saddle points even prior to the occurrence of the saddle-node bifurcation. When an unstable periodic orbit, from either the surrounding chaotic saddle or the gap regions in figure 6.7(b), approaches the vicinity of the location of these saddle points (cross), it is decelerated and spends more time in the regions shown in figures 6.7(c) and 6.7(d). In other words, all orbits of the chaotic attractor mimic (synchronize with) these period-1 unstable periodic orbits (M) when they come to their neighborhood (Kaplan 1993). This is the origin of the laminar phases seen in type-I intermittency of figure 6.6(a), which can also be explained in terms of phase synchronization of the unstable periodic orbits (Pikovsky et al. 1997; Pazo, Zaks and Kurths 2003; Pikovsky, Rosenblum and Kurths 2003). When a chaotic orbit moves away from the regions shown in figures 6.7(c) and 6.7(d), the orbit becomes desynchronized with respect to the unstable periodic orbit (M) created by the saddle-node bifurcation, which is manifested by the bursting phases in type-I intermittency of figure 6.6(a).

6.5 Attractor Merging Crisis

We study next what happens to the chaotic attractors at the end of the periodic window at $a_{MC} = 0.98765$. Chapter 5 showed that at a_{MC} an attractor merging crisis occurs due to the collision of two coexisting weakly chaotic attractors CA_1 and CA_2 with a pair of mediating unstable periodic orbits of period-3 and their associated manifold, which coincides with the boundary of the basins of attraction dividing the two weakly chaotic attractors. As the result of this crisis, two small chaotic attractors combine to form a single large chaotic attractor to the right of a_{MC}. Figures 6.8(a) and 6.8(b) are the enlargements, respectively, of the two rectangular regions of figure 6.4(a), showing the surrounding chaotic saddle SCS (black) and its stable manifold (the gray regions), the pair of weakly chaotic attractors CA_1/CA_2 (thin line), and the pair of period-3 mediating saddles (cross). The stable manifold of the

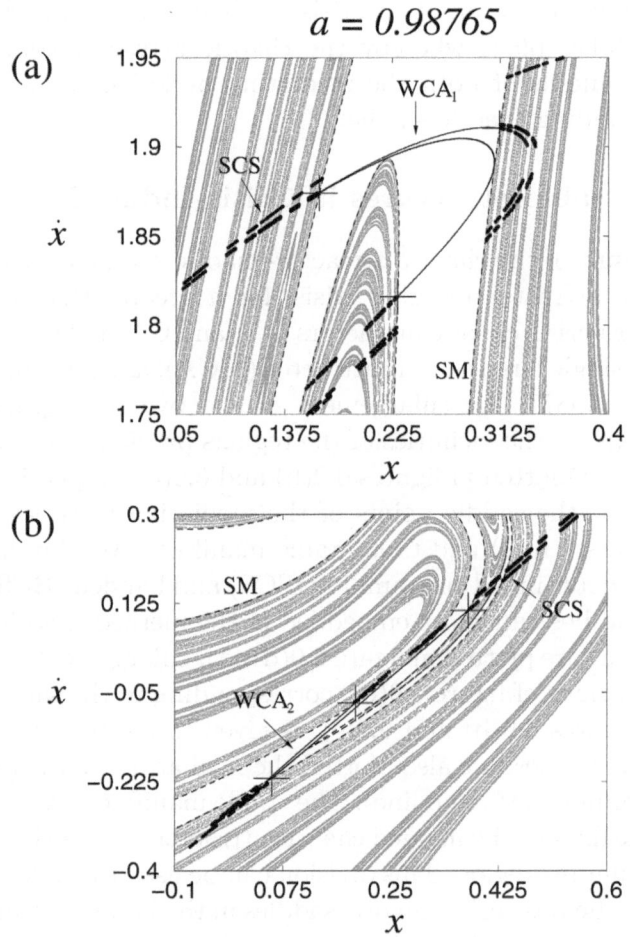

Fig. 6.8. Attractor merging crisis. Chaotic attractor-chaotic saddle collision at the attractor merging crisis for $a = 0.98765$. (a) Poincaré maps of the weak chaotic attractor WCA$_1$ (thin line), the surrounding chaotic saddle SCS (black), the stable manifold of SCS (gray), the mediating period-3 saddle (cross) and its associated stable manifold SM (dashed line); (b) Poincaré maps of the weak chaotic attractor WCA$_2$ (thin line), the surrounding chaotic saddle SCS (dark line), the stable manifold of SCS (gray), the mediating period-3 saddle (cross) and its associated stable manifold SM (dashed line). (a) and (b) correspond to the two rectangular regions indicated in figure 6.4(a).

mediating saddle is indicated by the dashed lines which separates the surrounding region occupied by the surrounding chaotic saddle from the band region occupied by the weakly chaotic attractors. Figure 6.8

reveals that at the onset of crisis, a chaotic attractor-chaotic saddle collision takes place whereby the chaotic attractor collides with the stable manifolds of both the mediating period-3 periodic saddle and the surrounding chaotic saddle.

6.6 Unstable Structures in Crisis-Induced Intermittency

As the result of the chaotic attractor-chaotic saddle collision at the onset of the attractor merging crisis, for a greater than a_{MC}, the two banded pre-crisis chaotic attractors CA_1 and CA_2 in figure 6.4(a) merge to form a single large chaotic attractor (MCA), as shown in figure 6.9(a), for $a = 0.9877$. An enlargement of the two rectangular regions of figure 6.9(a) in the vicinities of the regions previously occupied by CA_1 and CA_2 are plotted in figures 6.9(b) and 6.9(c), respectively, where we also plotted the saddle points of the mediating period-3 unstable periodic orbits (cross) and their stable manifold SM (thin line). The numerically determined surrounding (SCS) and banded (BCS_1 and BCS_2) chaotic saddles which are embedded in the merged chaotic attractor of figure 6.9(a) are plotted in figure 6.9(d). An enlargement of the two rectangular regions of figure 6.9(d), corresponding to the same regions covered by figures 6.9(b) and 6.9(c), is given in figures 6.9(e) and 6.9(f), respectively, where we also plotted the mediating saddle (cross) and its stable manifold SM (thin line). The stable manifold (SM) of the mediating saddle divides the merged chaotic attractor into the band region and the surrounding region. This division can be used to guide the numerical finding of the post-crisis chaotic saddles in the band and surrounding regions, respectively (Rempel et al. 2004a,b). It is evident from figure 6.9 that the banded chaotic saddles BCS_1 and BCS_2 (black) are located in the band regions previously occupied by the pre-crisis weakly chaotic attractors, since they are in fact converted from these two banded chaotic attractors at a_{MC}. BCS_1 and BCS_2 are found by a straddle orbit that never leaves the banded regions. Similarly, the surrounding chaotic saddle SCS (gray) is found by a straddle orbit that never enters the band regions.

It follows from the previous analysis that two nonattracting sets consisting of the surrounding chaotic saddle (SCS) and a pair of banded chaotic saddles (BCS_1 and BCS_2) are embedded in the post-crisis merged chaotic attractor (MCA), as shown in figure 6.9(d). Actually, the merged chaotic attractor is larger than the union of the surrounding and banded chaotic saddles, since the gaps in the post-crisis chaotic

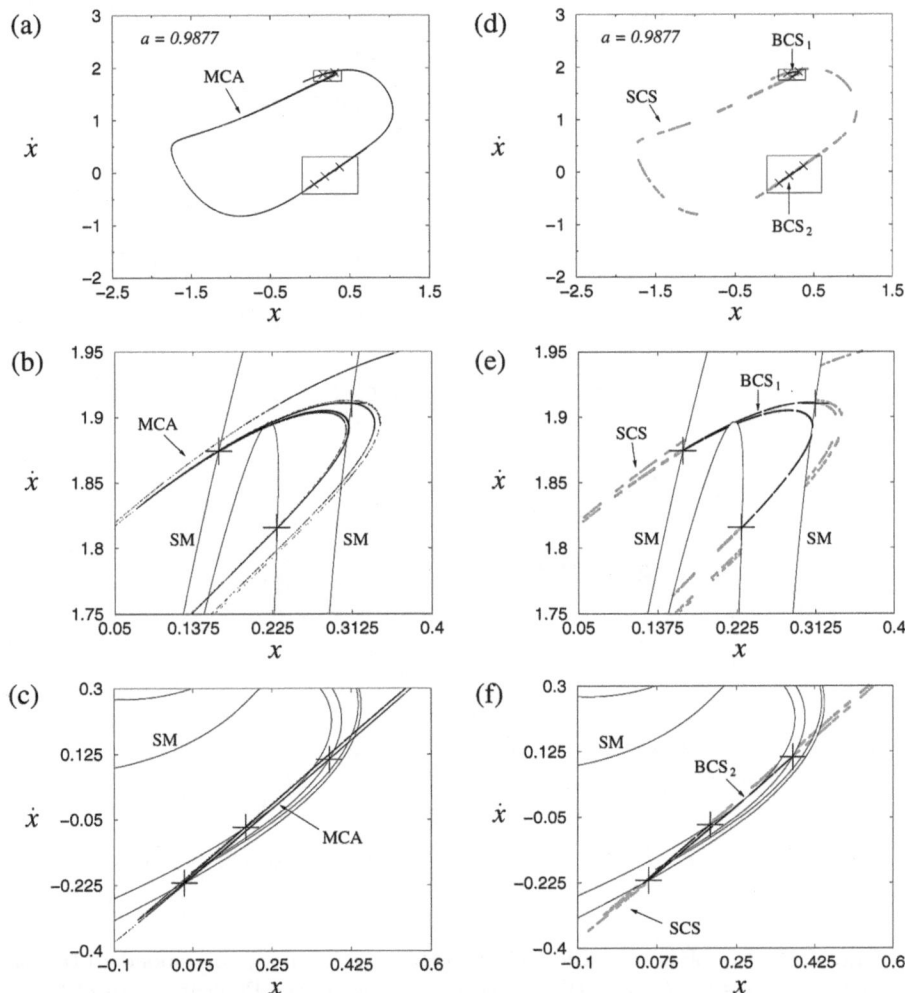

Fig. 6.9. Chaotic attractor and chaotic saddle in crisis-induced intermittency. (a) Poincaré map of the merged chaotic attractor (MCA) at $a = 0.9877$; (b) and (c) are enlargements of the two rectangular regions indicated in (a) showing the merged chaotic attractor (MCA), the period-3 mediating saddle (cross) and its associated stable manifold (SM); (d) Poincaré maps of the surrounding chaotic saddle SCS (gray) and the pair of banded chaotic saddles BCS$_1$ and BCS$_2$ (black) at $a = 0.9877$; (e) and (f) are enlargements of the two rectangular regions indicated in (d) showing the surrounding chaotic saddle (SCS), the banded chaotic saddles BCS$_1$/BCS$_2$, the period-3 mediating saddle (cross) and its associated stable manifold (SM).

saddles indicated in figures 6.9(d), 6.9(e) and 6.9(f) are not empty. They are densely filled by uncountably many unstable periodic orbits created by an explosion right after the onset of attractor merging crisis

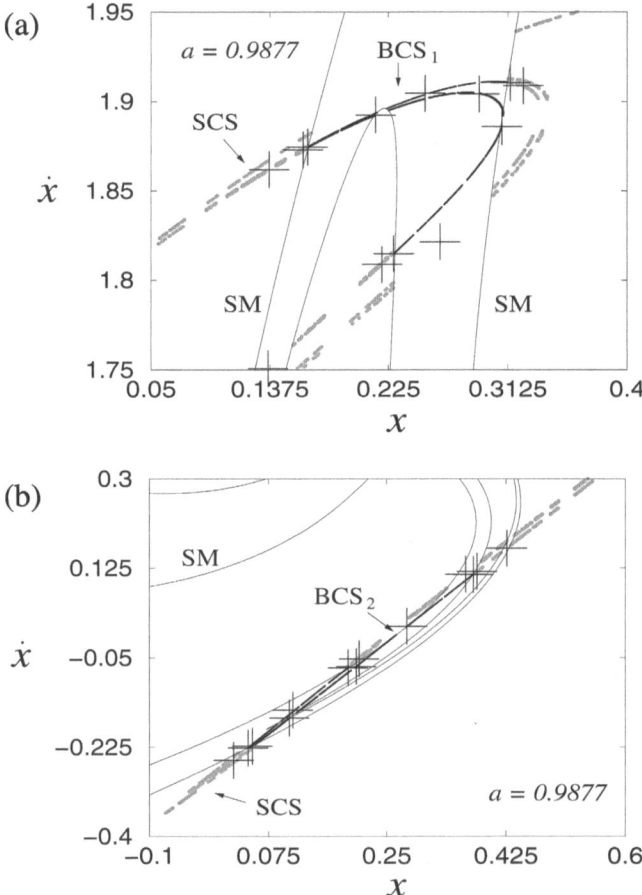

Fig. 6.10. Coupling unstable periodic orbits. (a) and (b) correspond to the same plots of figures 6.9 (e) and (f), respectively, showing the Poincaré maps of the surrounding chaotic saddle SCS (gray) and the pair of banded chaotic saddles BCS_1 and BCS_2 (black) at $a = 0.9877$, the stable manifold (SM) of the period-3 mediating saddles (not shown), and the pair of period-13 coupling unstable periodic orbits (cross) located in the gap regions of the surrounding and banded chaotic saddles.

(Szabó et al. 2000, Robert et al. 2000). This set of unstable periodic orbits within gaps, called coupling orbits with components in both band and surrounding regions, are responsible for the coupling between these two regions. Before crisis, for a less than a_{MC}, trajectories on the banded chaotic attractor never abandon the band region. For a slightly greater than a_{MC} a trajectory started in the band region can stay in that region for a finite duration of time, after which it crosses the stable

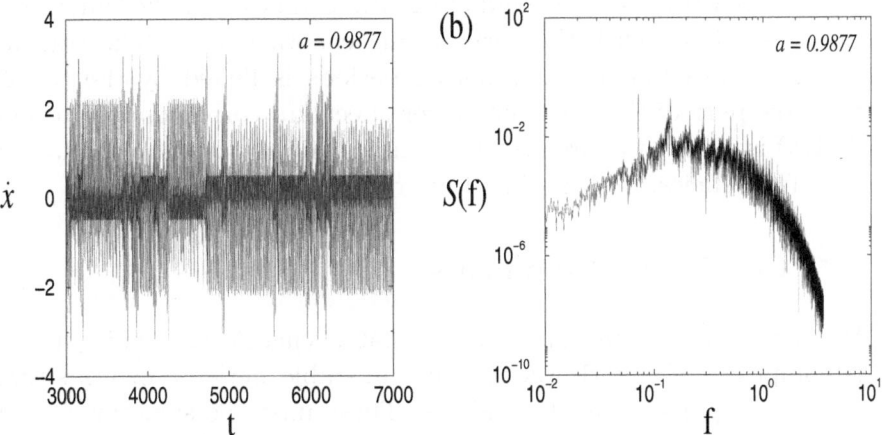

Fig. 6.11. Time series and power spectrum of crisis-induced intermittency. (a) Time series of crisis-induced intermittency, $\dot{x}$ as a function of time t, for $a = 0.9877$; (b) the power spectrum, $|\dot{x}|^2$ as a function of the frequency f, of the time series of (a).

manifold (SM) and escapes into the surrounding region. Once inside the surrounding region, the trajectory moves to the neighborhood of the surrounding chaotic saddle (SCS). After some time, the trajectory is injected back to the band region. This process of switching between the band and surrounding regions repeats intermittently. Hence, the coupling orbits located in the gaps of both banded and surrounding chaotic saddles link the trajectory from one region to the other; in principle, each switching may involve different coupling orbits. Right after crisis, the coupling orbits created by the explosion have very long period with the period approaching infinity as $a \to a_{MC}$ from above (Szabó et al. 2000). In that case, it is more difficult to numerically find a coupling orbit. However, as the control parameter a is increased further away from the crisis point a_{MC}, shorter coupling orbits are created. Figures 6.10(a) and 6.10(b) show a pair of period-13 coupling unstable periodic orbits numerically found in the same regions of figures 6.9(e) and 6.9(f) using the Newton algorithm (Curry 1979; Rempel et al. 2004a). Note that in figure 6.10 the Poincaré points of the coupling unstable periodic orbits are in fact located in the gaps of both banded and surrounding chaotic saddles.

Figure 6.11(a) shows the time series of crisis-induced intermittency, corresponding to the same control parameter of figures 6.9 and 6.10. This time series alternates episodically between the laminar periods associated with the two banded chaotic saddles and the bursting periods associated with the surrounding chaotic saddle. The transition between the laminar and bursting periods is linked by the coupling unstable periodic orbits. The power spectrum corresponding to figure 6.11(a) is given in figure 6.11(b), which exhibits power-law behavior at high frequencies, typical of real intermittent financial data.

6.7 Concluding Comments

We demonstrated in this chapter that a chaotic economic attractor is composed of chaotic saddles and unstable periodic orbits situated within the gaps of chaotic saddles. These unstable structures are the origin of intermittency in nonlinear economic models. In type-I economic intermittency, the laminar phases associated with low-level fluctuations of economic variables is a result of a phase synchronization of a chaotic orbit with the unstable periodic orbit created at the saddle-node bifurcation, whereas the bursting phases related to high-level fluctuations of economic variables are an indication that a chaotic orbit is far away from the unstable periodic orbit created at the saddle-node bifurcation. We showed that the attractor merging crisis in complex economic systems is due to a chaotic attractor-chaotic saddle collision, whereby two weakly chaotic attractors combine to form a large chaotic attractor. After the crisis, the pair of pre-crisis weakly chaotic attractors are converted into a pair of banded chaotic saddles. The post-crisis chaotic attractor is composed of the surrounding chaotic saddle, two banded chaotic saddles and coupling unstable periodic orbits in the gap regions which act as the link between the surrounding chaotic saddle and the banded chaotic saddles. In the time series of crisis-induced intermittency seen in figure 6.11(a), the laminar phases indicate that a chaotic orbit is in the region of the banded chaotic saddles, whereas the bursting phases indicate that a chaotic orbit is in the region of the surrounding chaotic saddle; the laminar and bursting phases are connected by the coupling unstable periodic orbits which have components in the gap regions of both surrounding chaotic saddle and banded chaotic saddles. Characteristic intermittency time, which measures the average duration of the laminar phases of either type-I or crisis-induced economic intermittency, can be calculated from the numerically simulated time series. This result can be useful for forecasting the turning point from bust to boom phases in business cycles.

7

Conclusion

In this monograph, we adopted a forced oscillator model of nonlinear economic cycles as a prototype to model the fundamental dynamical behaviors of a complex economic system, which exhibits multistability (coexistence of attractors), multiscale (power-law dependence on frequency), and coexistence of regularity and irregularity (order and chaos). It is important to point out that although we have selected the van der Pol model for its mathematical simplicity and its wide interest in economics, in view of the universal mathematical properties of nonlinear dynamical systems, the dynamical characteristics investigated in this simple model is actually applicable to other more sophisticated economic scenarios. We succeeded in characterizing the anatomy of a complex economic system by classifying its structure and dynamics. In terms of the system structure, our analysis shows that a complex economic system is composed of a hierarchy of stable and unstable structures, namely, stable and unstable manifolds of a fixed point in the state space and in the Poincaré section, stable and unstable periodic orbits, stable and unstable manifolds of a chaotic saddle, stable (periodic) and unstable (chaotic) attractors. In particular, we showed that unstable periodic orbits are the building blocks of chaotic saddles and chaotic attractors; moreover, chaotic saddles are embedded in a chaotic attractor and are responsible for the transient motion preceding the convergence to an attractor (periodic or chaotic). In terms of the system dynamics our results show that, as the control parameters are varied, a complex economic system undergoes a variety of dynamic transitions which change its stability properties, namely, local bifurcations such as period-doubling bifurcation, saddle-node bifurcation and

Hopf-bifurcation, and global bifurcations such as boundary crisis, interior crisis and attractor merging crisis.

Economic systems are unstable by nature, dominated by instabilities driven by both endogenous and exogenous forces. This very unstable nature of economic dynamics is clearly manifested by the unstable structures, such as unstable periodic orbits and chaotic saddles, inherent in chaotic economic systems. Recently, there is a surge of interest on the relevance of these unstable structures in economic dynamics. Lorenz and Nusse (2002) reconsidered the Goodwin's nonlinear accelerator model with periodic investment outlays and used it as an economic example of the emergence of complex motion in nonlinear dynamical systems. They showed that in addition to chaotic attractors, this model can possess coexisting attracting periodic orbits or simple attractors, which imply the emergence of transient chaotic motion (chaotic saddles). They applied straddle methods to numerically analyze this model in order to detect compact invariant sets which are responsible for the complexity of the transient motion, and concluded that chaotic saddles are prevalent in nonlinear economic models. Ishiyama and Saiki (2005) numerically found many unstable periodic orbits embedded in a chaotic attractor in a Keynes-Goodwin type of macroeconomic growth cycle model of two countries with different fiscal policies. These unstable periodic orbits not only look similar in shape to the chaotic attractor, there is a correspondence between the unstable periodic orbits and the chaotic attractor in terms of their statistical properties such as means, variances, Lyapunov exponents and probability density functions. Each value related to labor share rates, employment ratios, expected inflation rates and the instability of the chaotic attractor is almost the same as those of the unstable periodic orbits. Their results indicate that both statistical and dynamical features of a chaotic attractor in complex economic systems are captured by just a few unstable periodic orbits, in agreement with the periodic orbit theory of dynamical systems of Auerbach et al. (1987) and Cvitanovic (1988). This monograph renders strong support for the conclusions, that unstable periodic orbits and chaotic saddles are essential elements of complex economic systems, of Lorenz and Nusse (2002) and Ishiyama and Saiki (2005).

We demonstrated that intermittency is an intrinsic behavior of a chaotic economic system by analyzing in detail two examples of economic intermittency due to a local or a global bifurcation, namely, type-I intermittency and crisis-induced intermittency, respectively. The former is generated by a saddle-node bifurcation, the latter is generated by a crisis phenomenon such as the attractor merging crisis. In type-I intermittency, an economic system is capable of keeping the memory of its ordered dynamics before the transition to chaos; the time series of economic variables alternates between periods of seemingly periodic and chaotic fluctuations. In crisis-induced intermittency, an economic system is able to maintain the memory of its weakly chaotic dynamics before the transition to strong chaos; the time series of economic variables alternates between periods of weakly and strongly chaotic fluctuations. These two examples of chaos-driven intermittency can reproduce a number of patterns, namely, persistence, recurrence, memory, regime switching and volatility clustering, which are present in the intermittent time series observed in business cycles and financial markets (Diebold and Rudebusch 1999). The robustness of the unstable periodic orbits which form the skeleton of chaotic attractors and chaotic saddles can explain persistence, recurrence and memory patterns in business and financial cycles. The episodic switching between different dynamic states of an intermittent chaotic system can explain regime switching observed in economic and financial time series. The phase synchronization of unstable periodic orbits can be responsible for the spikes in the turbulent bursts as well as the quiescent phases in the time series, thus providing an explanation for the volatility clustering in financial data. Hence, the techniques developed in this paper for characterizing the complex dynamics of economic systems can become powerful tools for pattern recognition and forecasting of business and financial cycles. For example, the anticipation of the turning points is fundamental for forecasting business-cycle recessions and recoveries for countries showing asymmetric cycle durations (Garcia-Ferrer and Queralt 1998). Modeling of intermittency in nonlinear economic cycles can provide an estimate of the average duration of the contractionary phases of economic cycles and predict the turning points to expansionary phases. The classical NBER model of leading economic indicators was built solely within a linear framework which is inadequate for predicting the complex behavior of business cycles. By combining the complex system approach (such as chaotic theory developed in this paper) and the intelligent system approach (such as neural network), a superior performance for forecasting business cycle can be obtained relative to the classical model (Jagric 2003).

The techniques developed in this monograph can be readily applied to the study of chaos and complexity in management systems such as logistics and supply chain management (Mosekilde and Larsen 1988; Sosnovtseva and Mosekilde 1997), organizational dynamics and strategic management (Senge 1990; Stacey 2000), public policy and public administration (Kiel 1994). In fact, economic dynamics is a result of complex interactions of economical, political, social, climate, environmental and technological systems. For example, Berry (2000) performed an eigenanalysis of macroeconomic rhythms in the inflation rate and the rate of economic growth for the United States from 1790 to 1995, and obtained strong evidence of mode-locking of (Kondratiev) long waves by geophysical cycles; he suggested that a geophysical pacemaker may control the periodic appearance of long-wave crises, which leads to the clustering of innovations that drive successive surge of technological change. Nonlinear models of solar cycles, climate, and ecological systems indicate that these natural systems exhibit chaotic behaviors. Chian et al. (2003) showed that the dynamical systems approach is a powerful tool to model the complex dynamics of space environment and the solar-terrestrial relation which have great impact on climate, technology and environment. Sandor, Walsh and Marques (2002) discussed the rationale and objectives for pilot greenhouse-gas-trading markets, such as the Chicago Climate Exchange, now under development around the world; these markets represent an initial step in resolving a fundamental problem in defining and implementing appropriate policy actions to address climate change. Numerical modeling based on complex systems approach may be useful for the development of these emissions-trading markets, by assisting society to better understand the complex coupled energy-climate-environment system and assist policymakers to identify and implement optimal policies for managing the risks related to climate change.

The sensitive dependence of a dynamical system on small variations of its parameters can be used to control the chaotic behavior of a system by applying a small perturbation (Ott, Grebogi and Yorke 1990), which can be useful for stabilizing economic systems and optimizing management policies. This idea is based on the fact that a chaotic attractor has embedded in it an infinite number of unstable periodic orbits, which provides the flexibility to choose the most desirable periodic orbit whereby a chaotic system can be stabilized by introducing a small perturbation to convert it from an unstable periodic orbit to a stable periodic orbit. Lai and Grebogi (1994) showed that chaotic transient can be converted into sustained chaos by feedback control. There is evidence of chaos control in laboratory and numerical experiments. For example, Schief et al. (1994) applied the chaos method to control brain dynamics and succeeded to increase the periodicity of the in vitro neuronal population behavior and showed that neuronal systems can be made less periodic by applying chaos anticontrol techniques. Lopes and Chian (1996) showed that chaos in a coupled three-wave system, resulting from period-doubling bifurcations and type-I intermittency, can be controlled by applying a small wave with appropriate amplitude and phase. Kopel (1997) used a model of evolutionary market to show how firms can improve their performance in terms of profits if the decision makers of the firms apply the targeting method to switch from a chaotic evolution to a desired regular path. Kaas (1998) used the chaos control technique to show that the government can in principle stabilize an unstable Walrasian equilibrium in a short time by varying income tax rates or government expenditures. Rosser (2001) suggested that chaotic dynamics may actually be a desirable outcome for the sustainability of global complex ecologic-economic systems affected by climate change, as long as the policy agents are able to implement environmental policies that keep the system dynamics within sustainable levels by directing the management efforts at the appropriate levels of ecologic-economic interactions.

In this monograph, we only considered economic systems which are of low-dimension and varying only in time, described by ordinary differential equations. In many areas of economics and management, we must deal with dynamical systems which are of high-dimension and varying both in space and time. For example, in a study of fishery management of a lake district, Carpenter and Brock (2004) concluded that because of the complex interactions of mobile people and multistable ecosystems, optimal policies and management regimes will be highly heterogeneous in space and fluid in time. Some recent papers have demonstrated that nonlinear phenomena such as chaotic saddles, crisis, type-I intermittency and crisis-induced intermittency, observed in low-dimensional dynamical systems appear also in high-dimensional spatiotemporal dynamical systems (Chian et al. 2002, 2003; He and Chian 2003, 2004; Rempel et al. 2004b; Rempel and Chian 2005). Hence, the techniques developed in this monograph can be used to model complex spatiotemporal economical and managerial systems described by partial differential equations.

In this monograph, we have only focused on the deterministic characteristics of an economic system. Note, however, that uncertainty always plays a role in the economy, therefore a real economic system consists of both deterministic and stochastic dynamics (Hommes 2004). Barnett and Serletis (2000) reviewed the literature on the efficient markets hypothesis and chaos, and contrasted the martingale behavior of asset prices to nonlinear chaotic dynamics; in addition, they discussed the difficulty of distinguishing between probabilistic and deterministic behaviors in asset prices. Dhamala, Lai and Kostelich (2000) developed strategies to detect unstable periodic orbits from transient chaotic time series, in the presence and in the absence of noise, by examining recurrence times of trajectories in the vector space reconstructed from an ensemble of such time series, which can be useful for extracting unstable periodic orbits in intermittent economic and financial data. Small and Tse (2003) addressed the question of how to detect determinism in financial time series by examining daily returns from three financial indicators: the Dow Jones Industrial Average, the London gold fixings, and the U.S. dollar to Japenese Yen exchange rates; for each data set they applied surrogate data methods and nonlinearity tests to quantify determinism over a range of time scales, and found that all three time series are distinct from linear noise or conditional heteroskedastic models; they concluded that there exists detectable deterministic nonlinearity in real financial time series that can potentially be exploited for forecasting of financial markets.

In conclusion, characterization of nonlinear dynamical properties of economical time series obtained via numerical modeling may be the first step to understand the complex behavior of economic systems. Many of the traditional techniques being used by economists for modeling economic dynamics are based on linear approaches which are only valid near the equilibrium, and many of the tools being used by the investment professionals are based on the assumption that the asset returns have Gaussian distribution. In reality, the economic dynamics is often highly nonlinear and far away from the equilibrium, and the asset returns are usually intermittent with typically non-Gaussian distributions. The application of the complex systems approach developed in this paper to economic modeling and forecasting can improve decision making and policy planning, with positive impacts to the management of economic systems.

List of Figures

Bibliography

Alligood, K. T., Sauer, T. D., and Yorke, J. A. (1996). *Chaos: An Introduction to Dynamical Systems*. New York: Springer-Verlag.

Auerbach, D., Cvitanovic, P., Eckmann, J.-P., Gunaratne, G., and Procaccia, I. (1987). "Exploring chaotic motion through periodic orbits". *Physical Review Letters*. Vol. 58, pp. 2387-2389.

Bajo-Rubio, O., Fernandez-Rodriguez, F., and Soscilla-Rivero, S. (1992). "Chaotic behavior in exchange-rate series: first results for the Peseta-U.S. Dollar case". *Economics Letters*. Vol. 39, pp. 207-211.

Barnett, W. A., and Chen, P. (1988). "The aggregation-theoretic monetary aggregates are chaotic and have strange attractors: an econometric application of mathermatical chaos". In Barnett, W. A., Berndit, E. R., and White, H. (eds.). *Dynamic Econometric Modeling*. Cambridge: Cambridge University Press.

Barnett, W. A., and Serletis, A. (2000). "Martingales, nonlinearity, and chaos". *J. Economic Dynamics and Control*. Vol. 24, pp. 703-724.

Bautista, C. C. (2003). "Stock market volatility in Philippines". *Applied Economic Letters*. Vol. 10, pp. 315-318.

Belaire-Franch, J. (2004). "Testing for nonlinearity in an artificial financial market: a recurrence quantification approach". *J. Economic Behavior & Organization*. Vol. 54, pp. 483-494.

Benhabib, J. (1992). *Cycles and Chaos in Economic Equilibrium*. Princeton: Princeton University Press.

Berry, B. J. L. (2000). "A pacemaker for the long wave". *Technological Forecasting and Social Change*. Vol. 63, pp. 1-23.

Bischi, G. I., Stafanini, L., and Gardini, L. (1998). "Synchronization, intermittency and critical curves in a duopoly game". *Mathematics and Computers in Simulation*. Vol. 44, pp. 559-585.

Bordignon, S., and Lisi, F. (2001). "Predictive accuracy for chaotic economic models". *Economics Letters*. Vol. 70, pp. 51-58.

Borotto, F. A., Chian, A. C.-L., Hada, T., and Rempel, E. L. (2004). "Chaos in driven Alfvén systems: boundary and interior crises". *Physica D*. Vol. 194, pp. 275-282.

Borotto, F. A., Chian, A. C.-L., and Rempel, E. L. (2004). "Alfvén interior crisis". *Int. J. Bifurcation Chaos*. Vol. 14, pp. 2375-2380.

Brock, W. A. (1986). "Distinguishing random and deterministic systems: abridged version". *J. Economic Theory*. Vol. 40, pp. 168-195.

Brock, W. A., Hsieh, D. A., and LeBaron, B. (1991). *Nonlinear Dynamics, Chaos and Instability: Statistical Theory and Economic Evidence*. Cambridge: MIT Press.

Brock, W. A., and Hommes, C. H. (1997). "Rational route to randomness". *Econometrica*. Vol. 65, pp. 1059-1095.

Burns, A. F., and Mitchell, W. C. (1946). *Measuring Business Cycles*. New York: National Bureau of Economics Research.

Carpenter, S. R., and Brock, W. A. (2004). "Spatial complexity, resilence, and policy diversity: fishing on lake-rich landscapes". *Ecology and Society*, Vol. 9(1).

Chang, W. W., and Smyth, D. J. (1971). "The existence and persistence of cycles in a nonlinear model: Kaldor's 1940 model re-examined". *Rev. Economic Studies*. Vol. 38, pp. 37-44.

Chian, A. C.-L. (2001). "Nonlinear dynamics and chaos in macroeconomics". *Int. J. Theoretical and Applied Finance*. Vol. 3, pp. 601.

Chian, A. C.-L., Borotto, F. A., and Rempel, E. L. (2002). "Alfvén boundary crisis". *Int. J. Bifurcation Chaos*. Vol. 12, pp. 1653-1658.

Chian, A. C.-L., Rempel, E. L., Macau, E. E., Rosa, R. R., and Christiansen, F. (2002). "High-dimensional interior crisis in the Kuramoto-Sivashinsky equation". *Phys. Rev. E*. Vol. 65, 035203(R).

Chian, A. C.-L., Borotto, F. A., Rempel, E. L., Macau, E. E. N., Rosa, R. R., and Christiansen, F. (2003). "Dynamical systems approach to space environment turbulence". *Space Science Reviews*. Vol. 102, pp. 447-461.

Chian, A. C.-L., Rempel, E. L., Borotto, F. A., and Rogers, C. (2005). "Attractor merging crisis in chaotic business cycles". *Chaos, Solitons & Fractals*. Vol. 24, pp. 869-875.

Chian, A. C.-L., Rempel, E. L., and Rogers, C. (2006). "Complex economic dynamics: chaotic saddle, crisis and intermittency". *Chaos, Solitons & Fractals*. Vol. 29, pp. 1194-1218.

Chian, A. C.-L., Rempel, E. L., Borotto, F. A., and Rogers, C. (2006). "An example of intermittency in nonlinear economic cycles". *Applied Economics Letters*. Vol. 13, pp. 257-263.

Chian, A. C.-L., Rempel, E. L., and Rogers, C. (2007). "Crisis-induced intermittency in nonlinear economic cycles". *Applied Economics Letters*. Vol. 14, pp. 211-218.

Chiarella, C. (1988). "The cobweb – its instability and the onset of chaos". *Economic Modelling*. Vol. 5, pp. 377-384.

Chiarella, C. (1990). *The Elements of a Nonlinear Theory of Economic Dynamics, Series: Lecture Notes in Economics and Mathematical Systems*. Vol. 343. Berlin: Springer-Verlag.

Chu, P. C., Ivanov, L. M., Kantha, L. H., Melnishenko, O. V., and Poberezhny, Y. A. (2002). "Power law decay in model predictability skill". *Geophysical Research Letters*. Vol. 29, Art. No. 1748.

Curry, J. H. (1979). "An algorithm for finding closed orbits". *Global Theory of Dynamical Systems*. Eds. Nitecki, Z., and C. Robinson. Berlin: Springer-Verlag.

Cvitanovic, P. (1988). "Invariant measures of strange sets in terms of cycles". *Physical Review Letters*. Vol. 61, pp. 2729-2732.

Day, R. H. (1994). *Complex Economic Systems - Vol. 1: An Introduction to Dynamical Systems and Market Mechanisms*. Cambridge: MIT Press.

Day, R. H. (2000): *Complex Economic Dynamics - Vol. 2: An Introduction to Macroeconomic Dynamics*. Cambridge: MIT Press.

Day, R. H., and Pavlov, O. V. (2002). "Richard Goodwin's Keynesian cobweb: theme and variations". *Journal of Macroeconomics*. Vol. 24, pp. 1-15.

Dhamala, M., Lai, Y. C., and Kostelich, E. J. (2000). "Detecting unstable periodic orbits from transient chaotic time series". *Physical Review E*. Vol. 61, pp. 6485-6489.

Diebold, X. D., and Rudebusch, G. D. (1999). *Business Cycles - Durations, Dynamics and Forecasting*, Princeton University Press, Princeton.

Diebold, F. X., and Inoue, A. (2001). Long memory and regime switching, *Journal of Econometrics*. Vol. 105, pp. 131-159.

Ditto, W. L., Rauseo, R., Cawley, R., Grebogi, C., Hsu, G.-H., Kostelich, E. , Ott, E., Savage, H. T., Segnan, R., Spano, M. L., and Yorke, J. A. (1989). "Experimental observation of crisis-induced intermittency and its critical exponent". *Physical Review Letters*. Vol. 63, pp. 923-926.

Drepper, F. R., Engbert, R., and Stollenwerk, N. (1994). "Nonlinear time series analysis of empirical population dynamics". *Ecological Modelling*. Vol. 75, pp. 171-181.

Engelbrecht, J., and Kongas, K. (1995). "Driven oscillators in modelling of heart dynamics". *Applicable Analysis*. Vol. 57, pp. 119-144.

Faisst, H., and Eckhardt, B. (2003). "Traveling waves in pipe flow". *Physical Review Letters*. Vol. 91, 224502.

Frank, M. Z., Gencay, R., and Stengos, T. (1988). "International chaos". *European Economic Rev.*. Vol. 32, pp. 1569-1584.

Fernandez-Rodriguez, F., Sosvilla-Rivero, S., and Andrada-Felix, J. (1997). "Combining information in exchange rate forecasting: evidence from the SEM". *Applied Economics Letters*. Vol. 4, pp. 441-444.

Gabisch, G., and Lorenz, H. W. (1987). *Business Cycle Theory: a Survey of Methods and Concepts*. Berlin: Springer-Verlag.

Gandolfo, G. (1997). *Economic Dynamics*. Berlin: Springer-Verlag.

Garcia-Ferrer, A., and Queralt, R. A. (1998). "Using long-, medium-, and short-term trends to forecast turning points in the business cycles; some international evidence". *Studies in Nonlinear Economic Dynamics and Econometrics*. Vol. 3, pp. 79-105.

Ghashghaie, S., Breymannn, W., Peinke, J., Talkner, P., and Dodge, Y. (1996). "Turbulent cascades in foreign exchange markets". *Nature*. Vol. 381, pp. 767-770.

Gil-Alana, L. A. (2004). "The dynamics of the real exchange rates in Europe: a comparative study across countries using fractional integration". *Applied Economic Letters*. Vol. 11, pp. 429-432.

Goodwin, R. M. (1951). "The nonlinear accelerator and the persistence of business cycles". *Econometrica*. Vol. 19, pp. 1-17.

Goodwin, R. M. (1990). *Chaotic Economic Dynamics*. Oxford. Clarendon Press.

Granger, C. W. J., and Ding, Z. (1996). "Varieties of long memory models", *Journal of Econometrics*. Vol. 73, pp. 61-77.

Grebogi, C., Ott, E., and York, J. A. (1983). "Crises, sudden changes in chaotic attractors, and transient chaos". *Physica D*. Vol. 7, pp. 181-200.

Grebogi, C., Ott, E., Romerias, F., and Yorke, J. A. (1987). "Critical exponents for crisis-induced intermittency". *Phys. Rev. A*. Vol. 36, pp. 5365-5380.

Haxholdt, C., Kampmann, C., Mosekilde, E., and Sterman, J. D. (1995). "Mode-locking and entrainment of endogenous economic cycles". *System Dynamics Review*. Vol. 11, pp. 177-198.

Hayashhi, H., Ishizuka, S., and Hirakawa, K. (1983). "Transition to chaos via intermittency in the Onchidium Pacemaker Neuron". *Physics Letters A*. Vol. 98, pp. 474-476.

He, K. F., and Chian, A. C.-L. (2003). "On-off collective imperfect phase synchronization and bursts in wave energy in a turbulent state". *Physical Review Letters*. Vol. 91, 034102.

He, K. F., and Chian, A. C.-L. (2004). "Critical dynamic events at the crisis of transition to spatiotemporal chaos". *Physical Review E*. Vol. 69, 026207.

Hicks, J. R. (1950). *A Contribution to the Theory of the Trade Cycles*. Oxford: Clarendon Press.

Hilborn, R. C. (1994). *Chaos and Nonlinear Dynamics: An Introduction for Scientists and Engineers*. New York: Oxford University Press.

Hommes, C. H. (2004): "Economic dynamics.," in: A. Scott, ed., *Encyclopedia of Nonlinear Science*. London: Routledge, pp. 245-248.

Hsu, G.-H., Ott, E., and Grebogi, C. (1988). "Strange saddles and the dimensions of their invariant manifolds". *Physics Letters A*. Vol. 127, pp. 199-204.

Hughston, L. P., and Rafailidis, A. (2005). "A chaotic approach to interest rate modeling". *Finance and Stochastics*. Vol. 9, pp. 43-65.

Ishiyama, K., and Saiki, Y. (2005). "Unstable periodic orbits and chaotic economic growth". *Chaos, Solitons and Fractals*. Vol. 26, pp. 33-42.

Jagric, T. (2003). "A nonlinear approach to forecasting with leading economic indicators". *Studies in Nonlinear Dynamics and Econometrics*. Vol. 7 (2), Article 4.

Jánosi, I. M., Flepp, L., and Tél, T. (1994). "Exploring transient chaos in an NMR-laser experiment". *Physical Review Letters*. Vol. 73, pp. 529-532.

Kaas, L. (1998). "Stabilizing chaos in a dynamic macroeconomic model". *Journal of Economic Behavior and Organization*. Vol. 33, pp. 311-332.

Kaldor, N. (1940). "A model of the trade cycle". *Economic J*. Vol. 50, pp. 78-92.

Kalecki, M. (1937). "A theory of the business cycle". *Rev. Economic Studies*. Vol. 4, pp. 77-97.

Kantz, H., and Grassberger, P. (1985). "Repellers, semi-attractors and long-lived chaotic transients". *Physica D*. Vol. 17, pp. 75-86.

Kaplan, H. (1993). "Type-I intermittency for the Dénon-map family". *Physical Review E.* Vol. 48, pp. 1655-1669.

Keen, S. (1995). "Finance and economic breakdown: modeling Minsky's financial instability hypothesis". *Journal of Post Keynesian Economics.* Vol. 17, pp. 607-635.

Kholodilin, K. A. (2003). "US composite economic indicator with nonlinear dynamics and the data subject to structure breaks". *Applied Economic Letters.* Vol. 10, pp. 363-372.

Kiel, L. D. (1994). *Managing Chaos and Complexity in Government: A New Paradigm for Managing Change, Innovation and Organizational Renewal.* San Francisco: Jossey-Bass.

Kirikos, D. G. (2000). "Forecasting exchange rates out of sample: random walk vs Markov switching regimes". *Applied Economic Letters.* Vol. 7, pp. 133-136.

Konstantinou, K. I., and Lin, C. H. (2004). "Nonlinear time series analysis of tremor events recorded at Sangay volcano, Ecuador". *Pure and Applied Geophysics.* Vol. 161, pp. 145-163.

Kopel, M. (1997). "Improving the performance of an economic system: controlling chaos". *Journal of Evolutionary Economics.* Vol. 7, pp. 269-289.

Krawiecki, A., Holyst, J. A., and Helbing, D. (2002). "Volatility clustering and scaling for financial time series due to attractor bubbling". *Physical Review Letters.* Vol. 89, 158701.

Lai, Y.-C., and Grebogi, C. (1994). "Converting transient chaos into sustained chaos by feedback Control". *Physical Review E.* Vol. 49, pp. 1094-1098.

Lopes, S. R., and Chian, A. C.-L. (1996): "Controlling chaos in nonlinear three-wave coupling". *Physical Review E.* Vol. 54, pp. 170-174.

Lorenz, H.-W. (1987a). "Strange attractors in a multisector business cycle model". *J. Economic Behavior and Organization.* Vol. 8, pp. 397-411.

Lorenz, H.-W. (1987b). "Goodwin's nonlinear accelerator and chaotic motion". *J. Economics.* Vol. 47, p. 413-418.

Lorenz, H. W. (1993). *Nonlinear Dynamical Economics and Chaotic Motion.* Berlin: Springer Verlag.

Lorenz, H.W., and Nusse, H. E. (2002). "Chaotic attractors, chaotic saddles, and fractal basin boundaries: Goodwin's nonlinear accelerator model reconsidered." *Chaos, Solitons and Fractals.* Vol. 13, pp. 957-965.

Manneville, P., and Pomeau, Y. (1979). "Intermittency and the Lorenz model." *Physics Letters A*. Vol. 75, pp. 1-2.

Mantegna, R. N., and Stanley, H. E. (1995). "Scaling behaviour in the dynamics of an economic index". *Nature*. Vol. 376, pp. 46-49.

Mantegna, R. N., and Stanley, H. E. (1996). "Turbulence and financial markets". *Nature*. Vol. 383, pp. 587-588.

Mantegna, R. N., and Stanley, H. E. (2000). *An introduction to Econophysics: Correlations and Complexity in Finance*. Cambridge: Cambridge University Press.

Mattedi, A. P., Ramos, F. M., Rosa, R. R., and Mantegna, R. N. (2004). "Value-at-risk and Tsallis statistics: risk analysis of the aerospace sector." *Physica A*. Vol. 344, pp. 554-561.

Medio, A. (1992). *Chaotic Dynamics - Theory and Applications to Economics*. Cambridge: Cambridge University Press.

Mettin, R., Parlitz, U., and Lauterborn, W. (1993). "Bifurcation structure of the driven van der Pol oscillator." *Int. J. Bifurcation Chaos*. Vol. 3, pp. 1529-1555.

Mosekilde, E., and Larsen, E. R. (1988). "Deterministic chaos in the beer production-distribution system". *System Dynamics Reviews*. Vol. 4, pp. 131-147.

Mosekilde, E., Lasrsen, E. R., Sterman, J. D., and Thomsen, J. S. (1992). "Nonlinear mode-interaction in the macroeconomy". *Annals of Operations Research*. Vol. 37, pp. 185-215.

Muckley, C. (2004). "Empirical asset return distributions: is chaos the culprit?". *Applied Economic Letters*. Vol. 11, pp. 81-86.

Müller, U. A., Dacorogna, M. M., Olsen, R. B., Pictet, O. V., Schwarz, M., and Morgenegg, C. (1990). "Statistical study of foreign exchange rates, empirical evidence of a price change scaling law, and intraday analysis." *Journal of Banking and Finance*. Vol. 14, pp. 1189-1208.

Mullineux, A. W. (1990). *Business Cycles and Financial Crisis*. Hemstread: Harvester Wheatsheaf, Hemel.

Nusse, H. E., and Yorke, J. A. (1989). "A procedure for finding numerical trajectories on chaotic saddles." *Physica D*. Vol. 36, pp. 137-156.

Nusse, H. E., and Hommes, C. (1990). "Resolution of chaos with application to a modified Samuelson model". *J. Economic Dynamics and Control*. Vol. 14, pp. 1-19.

Ossendrijver, M., and Covas, E. (2003). "Crisis-induced intermittency due to attractor-widening in a buoyancy-driven solar dynamo". *Int. J. Bifurcation Chaos*. Vol. 13, pp. 2327-2333.

Ott, E. (1993). *Chaos in Dynamical Systems*. Cambridge: Cambridge University Press.

Ott, E., Grebogi, C., and Yorkes, J. A. (1990). "Controlling chaos." *Physical Review Letters*. Vol. 64, pp. 1196-1199.

Parker, T. S., and Chua, L. O. (1989). *Practical Numerical Algorithms for Chaotic Systems*. New York: Springer-Verlag.

Parlitz, U., and Lauterborn, W. (1987). "Period-doubling cascades and devil's staircases of the driven van der Pol oscillator." *Physical Review A*. Vol. 36, pp. 1428-1434.

Pazo, D., Zaks, M. A., and Kurths, J. (2003). "Role of unstable periodic orbits in phase and lag synchronization between coupled chaotic oscillators." *Chaos*. Vol. 13, pp. 309-318.

Perez-Munuzuri, V., and Gelpi, I. R. (2000). "Application of nonlinear forecasting techniques for meteorological modeling". *Annales Geophysicae*. Vol. 18, pp. 1349-1359.

Pikovsky, A., Zaks, M., Rosenblum, M., Osipov, G., and Kurths, J. (1997). "Phase synchronization of chaotic oscillations in terms of periodic orbits." *Chaos*. Vol. 7, pp. 680-687.

Pikovsky, A., Rosenblum, M., and Kurths, J. (2003). *Synchronization: A Universal Concept in Nonlinear Sciences*. Cambridge: Cambridge University Press.

Puu, T. (1989). *Nonlinear Economic Dynamics*. Berlin: Springer-Verlag.

Puu, T. (1991). "Chaos in duopoly pricing". *Chaos, Solitons & Fractals*. Vol. 1, pp. 573-581.

Puu, T., and Sushko, I. (2004). "A business cycle model with cubic nonlinearity". *Chaos, Solitons, & Fractals*. Vol. 19, pp. 597-612.

Rasmussen, S., Mosekilde, E., and Sterman, J. D. (1985). "Bifurcations and chaotic behavior in a simplified model of the economic long wave". *Systems Dynamics Reviews*. Vol. 1, pp. 92-110.

Rempel, E. L., Chian, A. C.-L., Macau, E. E. N., and Rosa, R. R. (2004a). "Analysis of chaotic saddles in low-dimensional dynamical systems: the derivative nonlinear Schrödinger equation". *Physica D*. Vol. 199, pp. 407-424.

Rempel, E. L., Chian, A. C.-L., Macau, E. E. N. , and Rosa, R. R. (2004b). "Analysis of chaotic saddles in high-dimensional dynamical systems: the Kuramoto-Sivashinsky equation." *Chaos*. Vol. 14, pp. 545-556.

Rempel, E. L., and Chian, A. C.-L. (2005). "Intermittency induced by attractor-merging crisis in the Kuramoto-Sivashinsky equation." *Physical Review E*. Vol. 71, 016203.

Resende, M., and Teixeira, N. (2002). "Permanent structural change in the Brazilian economy and long memory: a stock market perspective". *Applied Economic Letters*. Vol. 9, pp. 373-375.

Robert, C., Alligood, K. T., Ott, E., and Yorke, J. A. (2000). "Explosions of chaotic sets." *Physica D*. Vol. 144, pp. 44-61.

Rosser, J. B. (1991). *From Catastrophy to Chaos: a General Theory of Economic Discontinuities*. Boston: Kluwer Academic Publishers.

Rosser, J. B. (2001). "Complex ecologic-economic dynamics and environmental policy." *Ecological Economics*. Vol. 37, pp. 23-37.

Samuelson, P. A. (1939). "Interactions between the multiplier analysis and the principle of acceleration". *Rev. Economic Statistics*. Vol. 21, pp. 75-78.

Sandor, R., Walsh, M., and Marques, R. (2002). "Greenhouse-gas-trading markets." *Philosophical Transactions of the Royal Society of London Series A – Mathematical, Physical and Engineering Sciences*. Vol. 360, pp. 1889-1900.

Sasakura, K. (1995). "Political economic chaos?". *J. Economic Behavior and Organization*. Vol. 27, pp. 213-221.

Sayers, C. L. (1989). "Chaos and the business cycles". In Krasner, S. (ed.). *The Ubiquity of Chaos*. Washington: American Association of Science Publications.

Scarth, W. M. (1996). *Macroeconomics: An Introduction to Advanced Methods*. Toronto: Dryden.

Schief, S. J., Jerger, K., Duong, D. H., Chang, T., Spano, M. L., and Ditto, W. L. (1994). "Controlling chaos in the brain." *Nature*. Vol. 370, pp. 615-620.

Scheinkman, J. A., and LeBaron, B. (1989). "Nonlinear dynamics and GNP data". In Barnett, W. A., Geweke, J., and Shell, K. (eds.). *Economic Complexity: Chaos, Sunspots, Bubbles, and Nonlinearity*, pp. 213-227. Cambridge: Cambridge University Press.

Schnader, M. H., and Stekler, H. O. (1998). "Sources of turning point for forecast errors". *Applied Economics Letters*. Vol. 5, pp. 519-521.

Shone, R. (2002). *Economic Dynamics: Phase Diagrams and their Economic Application*. Cambridge: Cambridge University Press.

Selover, D., Jensen, R., and Kroll, J. (2004). "Industrial sector mode-locking and business cycle formation". *Studies in Nonlinear Dynamics and Econometrics*. Vol. 7, Art. No. 2.

Senge, P. M. (1990). *The Fifth Discipline: The Art & Practice of the Learning organization*. Sydney: Random House.

Sengupta, J. K., and Sfeir, R.E. (1997). "Exchange rate instability: some empirical tests of temporal dynamics". *Applied Economics Letters*. Vol. 4, pp. 547-550.

Small, M, and Tse, C. K. (2003). "Determinism in financial time series," *Studies in Nonlinear Dynamics and Econometrics*. Vol. 7 (3), Article 5.

Soofi, A. S., and Cao, L., (1999). "Nonlinear deterministic forecasting of daily Peseta-Dollar exchange rate". *Economics Letters*. Vol. 62, pp. 175-180.

Sosnovtseva, O., and Mosekilde, E. (1997). "Torus destruction and chaos-chaos intermittency in a commodity distribution chain." *Int. J. Bifurcation Chaos*. Vol. 7, pp. 1225-1242.

Stacey, R. D. (2000). *Strategic Management & Organisational Dynamics: The Challenge of Complexity*. Essex: Pearson Education.

Strogatz, S. H. (1994). *Nonlinear Dynamics and Chaos*. Cambridge: Persus Books Publishing.

Stutzer, M. J. (1980). "Chaotic dynamics and bifurcation in a macro model". *J. Economic Dynamics and Control*. Vol. 2, pp. 353-376.

Szabó, K. G., and Tél, T. (1994a). "Thermodynamics of attractor enlargement." *Physical Review E*. Vol. 50, 1070-1082.

Szabó, K. G., and Tél, T. (1994b). "Transient chaos as the backbone of dynamics on strange attractors beyond crisis." *Physics Letters A*. Vol. 196, pp. 173-180.

Szabó, K. G., Lai, Y.-C., Tél, T. , and Grebogi, C. (2000). "Topological gap filling at crisis." *Physical Review E*. Vol. 61, pp. 5019-5032.

Szydlowski, N., Krawiec, A., and Tobola, J. (2001). "Nonlinear oscillations in business cycle model with time lags". *Chaos, Solitons, & Fractals*. Vol. 12, pp. 505-517.

Thomas, L., S. Reitz, and Samanidou, E. (Eds.) (2005). *Nonlinear Dynamics and Heterogeneous Interacting Agents, Series: Lecture Notes in Economics and Mathematical Systems*. Vol. 550. Berlin: Springer.

Van der Pol, B., and van der Mark, J. (1928). "The heartbeat considered as a relaxation osicllation, and an electric model of the heart". *Phil. Mag.*. Vol. 6, pp. 763-775.

Vassilicos, J. C. (1995). "Turbulence and intermittency". *Nature*. Vol. 374, pp. 408-409.

Vilasuso, J. (1996). "Changes in the duration of economic expansions and contractions in the United States". *Applied Economic Letters*. Vol. 3, pp. 803-806.

Wolf, A., Swift, J. B., Swinney, H. L., and Vastano, J. A. (1985). "Determining Lyapunov exponents from a time series". *Physica D*. Vol. 16, pp. 285-317.

Xu., D., Li, Z., Bishop, S. R., and Galvanetto, U. (2002). "Estimation of periodic-like motions of chaotic evolutions using detected unstable periodic patterns". *Pattern Recognition Letters*. Vol. 23, pp. 245-252.

Xu, J.-X., and Jiang, J. (1996). "The global bifurcation characteristics of the forced van der Pol oscillator." *Chaos, Solitons and Fractals*. Vol. 7, pp. 3-19.

Zhang, W. B. (1990). *Synergetic economics – Dynamics, Nonlinearity, Instability, Non-Equilibrium, Fluctuations and Chaos*. Berlin: Springer.

Ziehmann, C., Smith, L. A., and Kurths, J. (2000). "Localized Lyapunov exponents and the prediction of predictability". *Physics Letters A*. Vol. 271, pp. 237-251.

Lecture Notes in Economics and Mathematical Systems

For information about Vols. 1–506
please contact your bookseller or Springer-Verlag